SPECIALTY AND MINOR CROPS HANDBOOK

SECOND EDITION

Heritage University Library
3240 Fort Road
Toppenish, WA 98948

Specialty and Minor Crops Handbook

SECOND EDITION

Publication 3346
UNIVERSITY OF CALIFORNIA
DIVISION OF AGRICULTURE
AND NATURAL RESOURCES
1998

For information about ordering this publication, contact

University of California
Division of Agriculture and Natural Resources
Communication Services—Publications
6701 San Pablo Avenue, 2nd Floor
Oakland, California 94608-1239

Telephone 1-800-994-8849 or (510) 642-2431
FAX (510) 643-5470
http://danrcs.ucdavis.edu
e-mail inquiries to danrcs@ucdavis.edu

Publication 3346

Printed on Recycled Paper

ISBN 1-879906-38-4
Library of Congress Card Catalog No. 95-062444

©1998 by the Regents of the University of California
Division of Agriculture and Natural Resources

All rights reserved
No part of this publication may be reproduced, stored in a retrieval system, or transmitted, in any form or by any means, electronic, mechanical, photocopying, recording, or otherwise, without the written permission of the publisher and the authors.

Printed in the United States of America.

On the cover: Dennis Tamura of Blue Heron Farms, Watsonville, California participates in the Tasting of Summer Produce in Oakland. Photo © 1987 by Yvonne Savio, used by permission.

The following photographs are used by permission of their copyright holders: fava beans, page 58, and kohlrabi, page 75 (© Sakata Seed America, Inc.); chive, page 43, and prickly pear cactus, page 94 (© Yvonne Savio); European black currant, page 56, gooseberry, page 67, red and white currant, page 102 (© Bernadine Strik).

To simplify information, trade names of products have been used. No endorsement of named products is intended, nor is criticism implied of similar products that are not mentioned.

The University of California, in accordance with applicable Federal and State law and University policy, does not discriminate on the basis of race, color, national origin, religion, sex, disability, age, medical condition (cancer-related), ancestry, marital status, citizenship, sexual orientation, or status as a Vietnam-era veteran or special disabled veteran. The University also prohibits sexual harassment.

Inquiries regarding the University's nondiscrimination policies may be directed to the Affirmative Action Director, University of California, Agriculture and Natural Resources, 300 Lakeside Drive, 6th Floor, Oakland, CA 94612-3560 (510) 987-0096.

2m-rev-11/97-WJC/VFG

Contributing Authors

JAMES A. BEUTEL is Cooperative Extension Pomologist (retired), Department of Pomology, UC Davis.

MICK CANEVARI is Farm Advisor, UC Cooperative Extension, San Joaquin County.

MARITA CANTWELL DE TREJO is Cooperative Extension Vegetable Specialist, Department of Vegetable Crops, UC Davis.

JANET CAPRILE is Farm Advisor, UC Cooperative Extension, Contra Costa County.

LAWRENCE CLEMENT is County Director, UC Cooperative Extension, Yolo and Solano Counties.

WES FOSTER is a UC Cooperative Extension Master Gardener in Santa Clara County.

IMO FU is former student intern, Small Farm Center, UC Davis.

NANCY GARRISON is Horticultural Advisor, UC Cooperative Extension, Santa Clara County.

LUIS HERNANDEZ is former Visiting Professor, Department of Vegetable Crops, UC Davis.

GARY HICKMAN is Farm Advisor, UC Cooperative Extension, San Joaquin County.

CAROL HILLHOUSE is a Demonstration Garden Coordinator at the Student Experimental Farm, UC Davis, and has worked with the amaranth breeding program at the Rodale Research Institute in Kutztown, Pennsylvania.

PAUL HORTON is a small farmer.

JOHN INMAN is Farm Advisor (retired), UC Cooperative Extension, Monterey County.

HUNTER JOHNSON, JR., is Cooperative Extension Vegetable Crops Specialist Emeritus, UC Riverside.

C. TODD KENNEDY, an attorney at law in private practice and a Santa Clara Valley fruit grower, is former secretary and member of the Board of Directors of California Rare Fruit Growers, Inc.

STEVEN KOIKE is Plant Pathology Farm Advisor, UC Cooperative Extension, Monterey and Santa Cruz Counties.

DEMETRIOS G. KONTAXIS is Pest Management/Public Information Program Advisor Emeritus, UC Cooperative Extension, Contra Costa County.

KEITH MAYBERRY is Farm Advisor, UC Cooperative Extension, Imperial County.

MIKE MURRAY is Farm Advisor, UC Cooperative Extension, Colusa County.

CLAUDIA MYERS is Information Technology Manager, DANR Communication Services, UC Davis, and former Associate Director of the Small Farm Center.

TONYA NELSON is former student intern, Small Farm Center, UC Davis.

WARNER OROZCO is former Staff Research Associate, Department of Vegetable Crops, UC Davis.

ED PERRY is Farm Advisor, UC Cooperative Extension, Stanislaus County.

LEE REICH is a horticultural consultant and writer based in New Paltz, New York, and has served as a researcher for both Cornell University and the U.S. Department of Agriculture.

CURT ROBINSON is former Marketing Manager, University Extension, UC Davis.

VINCENT RUBATZKY is Cooperative Extension Vegetable Specialist Emeritus, Deparment of Vegetable Crops, UC Davis.

YVONNE SAVIO is UC Master Gardener and Program Manager for Common Ground Garden Program, UC Cooperative Extension, Los Angeles County.

TED SIBIA is Librarian in the Bio/Ag Department, Shields Library, UC Davis.

JESUS VALENCIA is Farm Advisor, UC Cooperative Extension, Stanislaus County.

LOUIE VALENZUELA is former Area Farm Advisor, UC Cooperative Extension, Santa Barbara, San Luis Obispo, and Ventura Counties.

MARK VAN HORN is Director, Student Experimental Farm, UC Davis.

DAVID VISHER is Program Representative, Department of Vegetable Crops, UC Davis.

RON VOSS is Cooperative Extension Vegetable Specialist, Department of Vegetable Crops, UC Davis, and former Director of the Small Farm Program.

PAUL VOSSEN is Farm Advisor, UC Cooperative Extension, Sonoma County.

CARRIE YOUNG is with Sacramento Valley Milling, Inc., Ordbend.

Contents

PREFACE, ACKNOWLEDGEMENTS 1

INTRODUCTION 3

SPECIALTY AND MINOR CROPS
- Adzuki Bean 5
- Anise, Sweet Alice 7
- Arugula, Roquette, Rocket Salad 9
- Asian Pear 11
- Baby Corn 15
- Basil 17
- Belgian Endive, French Endive, Witloof Chicory, Chicon 20
- Bitter Melon 22
- Bok Choy 25
- Bottle Gourd, Calabash Gourd, Cucuzzi 27
- Canola 29
- Caper 32
- Cardoon 34
- Celtuce, Asparagus Lettuce 36
- Chayote, Mirliton, Vegetable Pear 37
- Chinese Broccoli, Kailan, Gai-lohn, Chinese Kale 39
- Chinese Long Bean, Yard-Long Bean, Asparagus Bean 41
- Chive 43
- Cilantro, Chinese Parsley, Coriander 45
- Citron, Preserving Melon 47
- Collards 48
- Daikon, Lobok, Oriental or Chinese Radish 50
- Dill 52
- Endive, Escarole, Chicory 54
- European Black Currant 56
- Fava Bean 58
- Fennel, Sweet Anise 60
- Fresh Figs 63
- Garlic Chive, Chinese Chive, Gow Choy 65
- Gooseberry 67
- Japanese Bunching Onion, Welsh Onion, Multiplier Onion 69
- Jicama, Yam Bean 71
- Kiwano, African Horned Cucumber or Melon, Jelly Melon 74
- Kohlrabi, Stem Turnip 75
- Leek 77
- Lemongrass, Citronella Grass 80
- Marjoram, Sweet Marjoram, Knot Marjoram 82
- Mung Bean 84
- Nappa Cabbage, Chinese Cabbage, Celery Cabbage, Pe-tsai 86
- New Zealand Spinach 88
- Okra, Gumbo 89
- Oregano, Winter Marjoram, Wild Marjoram, Pot Marjoram 91
- Parsnip 93
- Prickly Pear Cactus 94
- Purslane, Verdolaga 97
- Quinoa 98
- Radicchio, Red Chicory 100
- Red Currant, White Currant 102
- Rosemary 104
- Sage 107
- Salsify, Oyster Plant 110
- Specialty Lettuce 112
- Specialty Mustard 114
- Specialty Tomatoes 116
- Sponge Gourd, Chinese Okra, Luffa 120
- Swiss Chard 123
- Tarragon 125
- Thyme 127
- Tomatillo 130
- Turnip 132
- Vegetable Amaranth 134
- Water Convolvulus, Chinese Water Spinach, Swamp Cabbage 136
- Wax Gourd, Ash Gourd, Winter Melon, Christmas Melon 138

GLOSSARY OF ASIAN VEGETABLES 141

ANNOTATED BIBLIOGRAPHY 145
- Introduction 145
- Bibliography 146
- Books for Your Shelf 176
- Books from the Library 177
- Useful Free Publications 178
- Recommended Journals 179
- Seed Sources 179

INDEX TO SCIENTIFIC AND COMMON NAMES 181

Preface

Twenty years ago, very few people in the United States had heard of daikon, and only a handful of us would have considered eating or producing this specialty crop. But today, our meat and potatoes diet has given way to a remarkably diverse array of exotic and specialty produce including daikon and others such as luffa and quinoa. As immigrant groups have settled in the United States and imported and cultivated the products integral to their cultures, they have initiated a dramatic change in consumption and supply patterns across cultural and demographic lines. Many smaller-scale family farmers have discovered that the production and marketing of these specialty crops for a broader base of consumers has provided a new marketing niche and improved revenues.

This second edition of the *Specialty and Minor Crops Handbook* provides detailed information that will help you make intelligent decisions about growing the specialty crops that are right for you. We have greatly expanded this edition to include 63 crop sheets and an updated bibliography with new information. We hope this second edition will continue to be a useful guide to growers—whether commercial producers or backyard gardeners—as you integrate new specialty items into your farming systems or gardens. We are committed to being, and to keeping you, on the cutting edge of information.

DESMOND JOLLY
Director, Small Farm Center
University of California, Davis

Acknowledgments

The following people have contributed to the first or second edition of the *Handbook* either by reviewing leaflets or providing photographs (or both). Without their contributions, the present *Handbook* could not have been possible.

Harry Agamalian, UC Cooperative Extension Farm Advisor, Monterey County

Patricia Allen, Agroecology Program, UC Santa Cruz

Aziz Baameur, UC Cooperative Extension Farm Advisor, Riverside County

Beth Benjamin, Shepherd's Garden Seeds, Felton

Georgeanne Brennan, Le Marché Seeds International, Dixon

Erin Chapman, formerly of the Small Farm Center, UC Davis

Roberta Cook, Marketing Specialist, Agricultural Economics Department, UC Davis

Debra Deis, Sakata Seed America, Inc., Salinas
Ken Dwelley, Dwelley Farms, Brentwood
Bill Fujimoto, Monterey Market, Berkeley
Chet Fukushima, Principal Photographer, UC DANR Communication Services
Bettie Furuta, Furuta/Associates, Fallbrook
Charlotte Glenn, Le Marché Seeds International, Dixon
Gena Good, Harris Moran Seed Co., Hayward
John Guerard, UC Cooperative Extension Farm Advisor, Kern County
Tim Hartz, Specialist, Vegetable Crops Department, UC Riverside
Janine Hasey, UC Cooperative Extension Farm Advisor, Sutter–Yuba Counties
Ann Hipolito, former student intern, Small Farm Center, UC Davis
Shirley Humphrey, Small Farm Center, UC Davis
Pedro Ilic, UC Cooperative Extension Farm Advisor, Fresno County
Robert Kotch, Seneca Foods Corp., Pittsford, New York
Wendy Krupnick, Shepherd's Garden Seeds, Felton
Thomas F. Leigh, Entomologist, USDA Cotton Research Station, Shafter
Joyce McClellan, Ornamental Edibles, San Jose
Susan McCue, Small Farm Center, UC Davis
Richard Molinar, UC Cooperative Extension Farm Advisor, Alameda County
Mike Orzolek, Extension Vegetable Specialist, Pennsylvania State University
Harold Otto, UC Cooperative Extension Farm Advisor, Orange County
Suzanne Paisley, Senior Photographer, UC DANR Communication Services
Mike Reid, Professor, Environmental Horticulture Department, UC Davis
Richard Smith, UC Cooperative Extension Farm Advisor, San Benito, Monterey, and Santa Cruz Counties
Bif Soper, Student Experimental Farm, UC Davis
Bernadine Strik, Oregon State University, Corvallis, Oregon
Dennis Tamura, Blue Heron Farms, Watsonville
Steve Temple, Specialist, Agronomy Department, UC Davis
Mas Yamaguchi, Professor (retired), Vegetable Crops Department, UC Davis
Carrie Young, Sacramento Valley Milling, Inc., Ordbend
Warren Weber, Star Route Farm, Bolinas
Christie Wyman, formerly of the Small Farm Center, UC Davis

•

A very special thanks to Claudia Myers, former associate director, Small Farms Center, UC Davis, who compiled information for both the first and second editions of the *Specialty and Minor Crops Handbook*.

Introduction

What are specialty and minor crops? Quite simply they are fruits, vegetables, and herbs that are either imported or grown on a limited scale in the United States. An item that is a specialty or minor crop in this country could be quite common in another part of the world.

More specifically, specialty crops can be defined as the following:

- *Unusual or exotic crops.* Many are European, Asian, or Latin American crops that have been introduced by immigrants to the United States who brought with them their native cuisine. Examples are Belgian endive, bitter melon, radicchio, cilantro, daikon, and tomatillo.

- *Unusual varieties of such common crops as melons, tomatoes, peppers, lettuce, and apples.* For example, lettuces are now available in numerous shapes, colors, and sizes with names such as Red Oakleaf, Rouge d'Hiver, and Black-seeded Simpson.

- *Miniature or baby vegetables.* According to *The Packer*, "Most baby vegetables are fully ripe miniature vegetables cultivated for perfection. Others are immature vegetables picked before fully grown." We've all seen baby corn, squash, and pumpkins, but many more items can be sold as baby vegetables.

- *High-quality produce.* Another important specialty area is based strictly on quality. Though quality is hard to define, it basically consists of freshness, flavor, and appearance. Almost any item can be considered a specialty product using this definition, if it is grown and handled with exceptional care. High-quality fruits, vegetables, and herbs are in demand and command high prices.

Please feel free to send additional information, corrections, or changes for the crops currently in the *Handbook* to the Small Farm Center at the address that follows. Also send ideas or information about new crops you think we should know about.

Jim Stephens of the Institute of Food and Agricultural Sciences at the University of Florida put out a manual, *Minor Vegetables,* that we used as a model early in the development of the *Handbook.* Several of our crop sheets are adapted for California from his book. Two other major sources of information are *World Vegetables* by Vincent Rubatzky and Mas Yamaguchi (Chapman and Hall, 1997) and *Hortus Third* (Cornell University, Macmillan Publishing Co., 1976). For almost all crops, the botanical names we give correspond to those in *Hortus Third,* except that we have used the newer family names with the "-aceae" suffix: for example, "Brassicaceae" rather than the older "Cruciferae."

The annotated bibliography was assembled by Ted Sibia, head of the Bio-Ag Department at Shields Library, UC Davis, working with Paul Horton, a small-scale family farmer. They worked independently of the Specialty Crops Workgroup. Patricia Allen (UC Santa Cruz Agroecology Program) and Christie Wyman (formerly of the Small Farm Center) contributed a "Glossary of Asian Vegetables," which should help English speakers sort out the names and characteristics of many crops new to the United States and now in increasing demand. Both sections fit well with the purpose of the *Handbook,* and so are included here.

A word of caution: The crop sheets in this handbook by no means constitute a definitive source of information. Before you decide to grow a specialty crop commercially, you need to do your homework. Frieda Caplan of Frieda's Finest Produce Specialties in Los Angeles recommends the following basic steps:

1. Determine what can grow on your soils, given your climate and water availability. Your local Cooperative Extension Farm Advisor and the Soil Conservation Service can help you assess your land and climate's potential.

2. Talk to a good seed company to find out what seed or plant stock is available and what varieties the company recommends.

3. Talk to marketers to see if there is a market for the crop.

4. Once you've done all of the above, experiment with the crop by planting a small plot or a few rows to get some experience.

5. Be sure to plan how the product is to be packed and shipped. Does it need to be cooled and/or cleaned? What other postharvest handling and standard packs are required?

We hope you find the *Handbook* useful.

Small Farm Center
University of California
Davis, CA 95616
http://www.sfc.ucdavis.edu/

Adzuki Bean

Phaseolus angularis **is a member of the Leguminosae (legume) family.
Varieties include Japanese Red, Chinese Red, Adzuki Express (Johnny's Selected Seeds),
Takara (Japanese import), and Minoka (Minnesota Agricultural Experiment Station).**

The adzuki bean plant grows 1 to 2 feet high, with leaves resembling those of Southern peas. Flowers are yellow and are followed by a cluster of smooth, cylindrical pods with seeds 2 to 3 times larger than the planted seeds. Seeds usually are dark red, but can be green, straw-colored, black-orange, or mottled. They are round with a protruding ridge (seed scar) on the side.

Other name. *Azuki* (Japanese).

Market Information

The U.S. market for adzuki beans is limited. Acreage is contracted in advance of planting. New domestic and overseas markets are currently being developed.

Current production and yield. Adzuki bean is a major crop in Japan and China and is cultivated in Korea, New Zealand, India, Taiwan, Thailand, and the Philippines. In the United States, adzuki beans have been grown in Florida, Minnesota, and California. Yields are the greatest in lighter soils with irrigation and good drainage. Research plots in Minnesota have yielded an average of 1,400 pounds per acre.

Use. In Japan, adzuki bean is the second most important dry bean crop. The beans are cooked with rice or used in confections. In the United States, the main adzuki product is bean sprouts. Young, tender pods can be harvested as snap beans and eaten like snow peas or cooked like common green beans. Puréed beans are eaten as a vegetable and in baked foods. They have a slightly sweeter flavor than other beans.

Dried adzuki beans require 1 hour of soaking before boiling. To purée adzuki beans, mix the mature, dried beans with minced garlic, a pinch of tumeric or Chinese mustard, and some grated ginger. The purée can be served as a hot vegetable, mixed in sour cream or yogurt, used as a salad spread, or stuffed into mushroom caps. *An,* a mixture of adzukis, sugar, and water, serves as a filling for bread, steamed breads or dumplings, and sweet cakes.

*Dried adzuki beans at the market.
(Photo: Suzanne Paisley)*

Nutrition. The beans are high in protein (25%) and easy to digest.

Culture

Climatic requirements. Seeds do well during frost-free periods with cool nights. The plant is reported to be somewhat drought resistant. Requirements for adzuki beans are similar to those for soybeans or other dry beans.

Propagation and care. Adzuki is a short-day plant that does not grow well in waterlogged soil. The seeds may be treated for fungi, insects, and bacteria before planting. In Minnesota, the best planting dates come in May and June. Bean plants emerge more slowly when the soil is 50° to 55°F. A good plant population is 105,000 plants (25 to 35 pounds of seed) per acre.

Plant seeds in rich, loamy soil, ½ to 1 inch deep. Plants should stand 2 to 3 inches apart. Recommended space between rows varies between 12 to 18 inches and 18 to 30 inches. Neutral to alkaline soils are required for maximum nitrogen fixation. Adzuki beans will fix nitrogen, but to do so they must be inoculated with a crop-specific Rhizobium strain. Fertilize seedlings when they are 4 to 5 inches tall, and again when the flowers start to form pods. Moisture should be ample and consistent.

Seed quality is important for vigor in young plants, since this crop competes poorly against weeds. Choose a location with light weed pressure, and rotary hoe 7 to 10 days after planting. Cultivate the beans when the primary leaves are fully developed and, if necessary, 10 to 20 days later.

Pests and diseases. White mold, bacterial stem rot, and other bean diseases may affect adzuki beans. A rotation program, furrow irrigation (rather than overhead), disease-free seed, and a spray program can help prevent these diseases. Most varieties are susceptible to aphid-borne viruses that attack legumes, including curly top virus.

Harvest and postharvest practices. To harvest as green beans, pick the adzuki pods when the beans are faintly outlined in the pod. Picking every 5 to 6 days is usually sufficient. In California, adzuki beans will mature in fewer than 120 days for use as dry beans.

Growers can cut and windrow in the morning to allow drydown and then collect seeds with a combine later in the day, or they can direct-combine the beans with a grain header or row crop headers. Adzuki pods shatter easily, especially if they are harvested late in the season or late in the day. To decrease losses, use slower speeds, open the concaves, and harvest only during appropriate hours. The entire plant can be harvested and stacked in a dry, well-ventilated place for drying. Complete drying occurs in 1 to 2 weeks. After drying, shell the beans and store them in refrigerated, air-tight containers.

Sources

Seed

Dr. Yoo Farm, P.O. Box 290, College Park, MD 20740

Hudson Seedsman, P.O. Box 1058, Redwood City, CA 94064

Johnny's Selected Seeds, 305 Foss Hill Road, Albion, ME 04910

Mellinger's Inc., 2340 S. Range Road, North Lima, OH 44452

Redwood City Seed Co., P.O. Box 361, Redwood City, CA 94064

Southern Exposure Seed Exchange, P.O. Box 158, North Garden, VA 22959

Sunrise Enterprises, P.O. Box 10058, Elmwood, CT 06110

Vermont Bean Seed Co., Garden Lane, Fair Haven, VT 05743

More information

Breene, William M., and Leland L. Hardman. 1987. Anatomy of a specialty crop — The adzuki bean experience. *Grain legumes as alternative crops,* pp. 67–76. The Center for Alternative Crops and Products, University of Minnesota.

Hardman, L. L., E. S. Oplinger, J. D. Doll, and S. M. Combs. 1989. Adzuki bean. *Alternative field crops manual.* Departments of Agronomy and Soil Science, Cooperative Extension Service and College of Agricultural and Life Sciences, University of Wisconsin, Madison, WI.

Harrington, Geri. 1978. *Grow your own Chinese vegetables.* Garden Way Publishing, Pownal, VT.

Robinson, R. G. 1980. Registration of Minoka adzuki bean (Reg. No. 19). *Crop Science.* Crop Science Society of America, Madison, WI. July/Aug 1980, 20(4):549.

Rubatzky, Vincent, and Mas Yamaguchi. 1997. *World vegetables, 2d ed.* Chapman and Hall, New York, NY.

Stephens, James. *Minor vegetables.* 1988. Cooperative Extension Bulletin SP-40, University of Florida, Gainesville, FL.

Whealy, Kent. 1988. *Garden seed inventory: 2d ed.* Seed Saver Publications, Decorah, IA.

Prepared by Tonya Nelson.

Anise, Sweet Alice

Pimpinella anisum is a member of the Apiaceae (parsley) family.

Anise is an annual herb that reaches a height of about 2 feet. Leaves and seeds are produced in large, loose clusters. Young leaves are broad, up to 1 inch wide, and resemble parsley, while the older leaves are feathery like dill. Small, yellowish white flowers form in rounded clusters with five petals, five stamens, and two styles. Seeds — really dried fruits — are about ⅛ inch long flattened, oval, downy, and gray-brown, with lengthwise ribs; they're produced in large, loose umbels on tall, rounded, grooved stalks. Anise is frequently confused with sweet anise (also known as Florence fennel and finnochio), but is really quite different. Anise is native to Egypt and the Mediterranean region.

Market Information

Current production and yield. Average yields in California range from 500 to 700 pounds of seed per acre.

Use. Today, anise is cultivated in Europe, India, Mexico, Russia, and the United States. The seed has been used for centuries to flavor pastries, candies, and beverages. The oil distilled from the seed may be used when the seed itself would have an undesirable appearance in the edible product. The oil is also used in medicines and in perfumes, soaps, and other toilet articles. The flavor of the leaves, seed, and oil is sweet, similar to licorice. The green leaves and stalks can be eaten raw, steamed, sautéed, or dried.

Culture

Climatic requirements. Anise requires a frost-free season of at least 120 days for growth. Uniform rainfall or irrigation throughout the growing season is essential because the plant is adversely affected by fluctuating soil moisture.

Propagation and care. The plant grows best in a light, fertile, sandy loam soil with good drainage. Plant the seed about ½ inch deep in rows 18 to 30 inches apart, about one or two seeds per inch. At this rate, 5 to 10 pounds of seed will plant 1 acre. Anise does not transplant easily. Germination will occur in 7 to 14 days at 70°F. Carefully thin seedlings to 6 or 12 inches apart; again, they do not transplant well. Growth is spindly, so hill-up the soil slightly to hold the new plants upright; additional support may be needed later.

Anise reaches a height of about 2 feet; the young leaves are as wide as 1 inch. (Photo: Marita Cantwell de Trejo)

While the seed in the umbel is still green, commercial growers either pull the plants out of the ground or cut the tops by hand. They then tie the harvested material into bundles and stack them in conical piles with the fruiting heads toward the center. The seed continue to ripen with little discoloration or shattering. After this curing, the crop is threshed.

Rust can be severe under California conditions.

Storing seeds in airtight containers will help retain flavoring properties for years. Oil distilled from seeds should be stored in a tightly sealed container away from excessive heat.

Sources

Seed

Abundant Life Seed Foundation, P.O. Box 772, Port Townsend, WA 98368

Bountiful Gardens, 5798 Ridgewood Road, Willits, CA 95490

Caprilands Herb Farm, 534 Silver Street, Coventry, CT 06238

Comstock, Ferre & Co., 263 Main Street, Wethersfield, CT 06109

De Giorgi Co., Inc., P.O. Box 413, Council Bluffs, IA 51502

Gurney's Seed & Nursery Co., Yankton, SD 57079

J. L. Hudson, Seedsman, P.O. Box 1058, Redwood City, CA 94064

Johnny's Selected Seeds, 299 Foss Hill Road, Albion, ME 04910

Le Jardin du Gourmet, P.O. Box 75, St. Johnsbury Center, VT 05863

Nichols Garden Nursery, 1190 North Pacific Highway, Albany, OR 97321

Otto Richter & Sons Ltd., Box 26, Goodwood, Ontario, Canada L0C 1A0

Pinetree Garden Seeds, Route 100, New Gloucester, ME 04260

Redwood City Seed Co., P.O. Box 361, Redwood City, CA 94064

Stokes Seeds Inc., P.O. Box 548, Buffalo, NY 14240

Taylor's Herb Gardens, Inc., 1535 Lone Oak Road, Vista, CA 92084

Thompson & Morgan, P.O. Box 1308, Jackson, NJ 08527

More information

Kowalchik, Claire, et al., eds. 1987. *Rodale's illustrated encyclopedia of herbs,* pp. 14–16. Rodale Press, Emmaus, PA.

Newcomb, Duane, and Karen Newcomb. 1989. *The complete vegetable gardener's sourcebook.* Prentice Hall Press, West Nyack, NY.

Organic Gardening Magazine. 1978. *The encyclopedia of organic gardening,* pp. 38–39. Rodale Press, Emmaus, PA.

Stephens, James. *Minor vegetables.* 1988. Cooperative Extension Bulletin SP-40, University of Florida, Gainesville, FL.

Williams, Louis. 1960. *Drug and condiment plants,* pp. 5–6. Agric. Handb. 172, USDA-ARS, Washington, DC.

Prepared by Yvonne Savio and Claudia Myers.

Arugula, Roquette, Rocket Salad

***Eruca sativa* is a member of the Brassicaceae (mustard) family.**

The edible leaves of arugula are characterized by a distinctive sharp, spicy, pungent, peppery flavor similar to that of mustard greens. It is a low-growing, 8- to 24-inch herbaceous annual with green, deeply cut compound leaves. Plants grow densely and develop white or yellowish blossoms with deep violet or reddish veins. Native to the Mediterranean region, arugula grows wild throughout southern Europe.

Market Information

Use. Ancient Romans and Egyptians considered arugula to be an aphrodisiac. The British cultivated it for centuries, and it was in the earliest gardens of New England. Today arugula is popular in Italy, France, Spain, Greece, and Egypt; its popularity is growing in the United States. In India, arugula is grown primarily for the oil that is obtained from the seeds; the leaves are not used.

Arugula may be harvested as an entire plant or as individual leaves cut from the plant. Raw in mixed salads, the leaves complement both bland butterhead lettuce and bitter chicories. The best time to use it raw in salads or in tomato dishes is when the serrated leaves are only 2 to 3 inches long. Arugula can be steamed, cooked as a potherb, or puréed and added to soups. You can freeze harvested leaves for later use, like spinach or other greens.

Culture

Climatic requirements. Arugula is a cool-season vegetable. In summer, the leaves will develop an unpleasantly strong flavor, and plants will bolt.

Propagation and care. For best results, plants should grow quickly and steadily. Leaf flavor gets stronger with warm weather and less irrigation. To ensure optimum growth and flavor, sow seeds early in cool weather in loose, well-composted soil. Keep the soil moist.

Sow the seed as early in spring as the soil can be worked, about ½ inch deep in rows 12 inches apart. When seedlings are 2 to 3 inches tall, thin plants to 6 or 9 inches apart. For a continuous crop, make sowings every few weeks. In mild winter areas, sow again in fall for winter harvest.

Arugula or rocket salad grows to a height of 8 to 24 inches. (Photo: Hunter Johnson)

Arugula is also an excellent late-season crop. Sown in early fall, it can withstand light frost. As the weather turns cold, protect the plants with cold frames or row covers.

Leaves will be ready to harvest 6 weeks after planting, when the plants are between 8 and 10 inches tall. At the peak of freshness, the leaves are dark green and somewhat smooth. A furry underside indicates toughness. Continuous harvest of young leaves will encourage further leaf production and a more prolific crop. The leaves become bitter after the plant has flowered, when the flat, open plant becomes leggy and will grow to a foot or more in height.

Harvesting is done by hand. The harvester cuts, bunches, and packs the crop into cartons in the field. Keep the leaves clean. Arugula is extremely perishable and needs gentle handling and rapid marketing.

Flea beetles can cause major problems as the weather warms.

Sources

Seed

NOTE: Seeds are widely available.

More information

Chandoha, Walter. 1984. Grow Italian greens. *Organic Gardening.* May 1984, pp. 80–84.

Halprin, Anne, ed. 1978. *Gourmet gardening,* pp. 189–91. Rodale Press, Emmaus, PA.

Mansour, N. S. 1990. *Rocket.* Vegetable Crop Recommendations, Oregon State University, Corvallis, OR.

Newcomb, Duane, and Karen Newcomb. 1989. *The complete vegetable gardener's sourcebook.* Prentice Hall Press, West Nyack, NY.

Organic Gardening Magazine. 1978. *The encyclopedia of organic gardening.* Rodale Press, Emmaus, PA.

Rubatzky, Vincent, and Mas Yamaguchi. 1997. *World vegetables, 2d ed.* Chapman and Hall, New York, NY.

Stephens, James. *Minor vegetables.* 1988. Cooperative Extension Bulletin SP-40, University of Florida, Gainesville, FL.

Whealy, Kent. 1988. *Garden seed inventory, 2d ed.* Seed Savers Publications, Decorah, IA.

Prepared by Yvonne Savio and Claudia Myers.

Asian Pear

All Asian pears grown today are selected seedlings or crosses made within the species *Pyrus serotina*, a member of the Rosaceae (rose) family.

The term "Asian pear" describes a large group of pear varieties having crisp, juicy fruit. When mature, the fruit are good to eat when harvested or for several months after picking if held in cold storage. The crisp texture of an Asian pear remains unchanged after picking or storage, unlike the flesh of European pears such as Bartlett or Comice.

There are three types of Asian pear: round or flat fruit with green to yellow skin; round or flat fruit with bronze-colored skin and a light bronze-russet; and pear-shaped fruit with green or russet skin.

Other names. Asian pears are called apple pears, salad pears, *Nashi* (Japanese for "pear"), Oriental, Chinese, or Japanese pears (*Nihonnashi*).

Ripe Asian pears have crisp, juicy flesh. (Photo: Suzanne Paisley)

Market Information

The present market consists of ethnic Asians in the western United States, Vancouver (Canada), and major cities in the United States. In the early 1980s demand stimulated new plantings, and most of these are now in production. Existing plantings probably will fill present demand, but more plantings may be necessary to fill potential demand. The future market includes consumers who want crisp, sweet pears that do not get soft and can be eaten as soon as purchased at the market. Some Asian pears are exported from California.

Current production and yield. Japan exports Asian pears to the United States in October and November. China and Korea also export to the United States and Canada. In recent years, plantings were made in New Zealand, Australia, Chile, France, and the eastern and southeastern United States.

About 4,000 to 5,000 acres have been planted in California, Oregon, and Washington. Production costs are about $2,500 to $3,000 per acre. Yields are lower than for Bartlett or Bosc pears because of the heavy thinning necessary to get the large sizes the market demands. Yields in the fifth to seventh years range from 200 to 500 packed boxes per acre. Trees aged 10 to 14 years may yield 800 to 1,000 packed boxes per acre of 30-, 40-, and 50-size fruit.

Culture

Climatic requirements. Asian pear trees such as 20th Century are about as winter hardy as Bosc pears, tolerating about –20°F, but are less hardy than Bartlett and Anjou. Asian pear rootstocks' tolerances for winter cold are 10°F for *P. calleryana,* 0° to –10°F for *P. betulaefolia,* and –30°F for *P. communis* and *P. serotina.*

Propagation and care. There is no standard accepted spacing for Asian pears on the West Coast. Plantings range from 7½ by 15 feet to 15 by 20 feet depending on soil, rootstock, and grower preference. In general, 200 trees per acre spaced 12 feet apart in rows with 17 to 18 feet between rows is a good pattern. Plantings of 145 to 200 trees per acre are recommended for vigorous selections and rootstocks.

Bloom period varies with the variety. Some early blooming Chinese varieties bloom 10 to 14 days before Bartlett. In the San Joaquin and Sacramento valleys of California, these early blooming varieties are at full bloom in early to mid-March. They are the first pears to bloom and are most subject to frost damage. Other early Japanese varieties flower at the same time as Anjou or Winter Nelis. A few late-flowering Japanese varieties reach full bloom with Bartletts.

In California, most Asian pear trees are trained in a vase shape. Head the nursery trees 25 to 30 inches high at planting, and select three or four main limbs the first year. Head these about 50 percent, leaving 12 to 24 inches of growth. This provides six to ten fairly low secondary limbs that are headed 30 to 36 inches long in the second dormant season. After fruit production starts (the third season), limbs are allowed to grow about 18 inches per year and then they are headed in the dormant season. If trees grow excessively, reduce pruning to encourage fruit spur development.

Pears are borne on spurs on 2- to 6-year-old wood, but the best sizes come from 1- to 3-year-old spurs on wood 1 to 2 inches in diameter.

Pollination. Asian pear varieties are partially self-fruitful, but better crops are set where two or more varieties are planted together. Cross-pollination usually gives better fruit size. Many varieties pollinate each other, but not all will do so. In Fresno and Tulare counties, 20th Century or Shinseiki are known to set good crops when planted alone in large, one-variety blocks. In areas with cooler temperatures at bloom-time, cross-pollination by European or Asian pear varieties will be necessary. Cross-pollinated fruit with seed tend to be larger and more uniformly round than inadequately pollinated fruit with few seeds.

There are no proven guidelines on the nearness of pollenizers or the use of bees in California. It is suggested that every four to eight rows of single-variety planting have a pollenizer row, or that growers plant a block or four to eight rows of a second variety adjacent to the first variety. Bees may be used at a density of one or two hives per acre. Too much pollination will require more thinning to achieve desired fruit sizes.

Fruit thinning. Asian pear varieties require heavy thinning, which is done manually since chemical thinning is not safe or reasonably effective. Some growers blossom-thin by hand. Most growers wait for fruit to set and then cut off all but one or two fruit per spur. This first fruit thinning is best done before the first codling moth spray, and can be accomplished by cutting the fruit off with clippers or small hand shears. A second follow-up thinning before the second codling moth spray is necessary for hard-to-size varieties and to remove pears not properly thinned the first time. For the best thinning, it usually takes two rounds to effectively leave no more than one fruit per spur; if spurs are close together, well-thinned fruit are spaced 4 to 6 inches apart. Thinning as close as 30 days before harvest can benefit size, but early thinning is essential for annual bearing and good fruit sizes. Thinning will require up to one-half hour per tree on younger trees and one hour or more on older trees.

Trees 4 to 5 years old size fruit easily if they have only 100 fruit per tree after thinning. Crop loads of 200-plus fruit per tree are common on 8- to 10-year-old trees. In Japan, 500 to 700 fruit are recommended on large bearing trees, yielding 70,000 fruit per acre.

Irrigation. Trees require frequent irrigation to produce large fruit sizes. High vigor, however, makes fireblight harder to control and makes fruit tender to pick.

Harvest and postharvest practices. In general, harvest in California runs from mid-July through September. In Washington and Japan, harvest time is from August to October. Harvest date is determined by color, sugar content, and fruit pressure. At least three color picks are necessary to get mature, quality fruit from most varieties in the Central Valley of California.

Fruit must be harvested carefully into padded picking buckets or boxes and handled gently in the packing house to minimize handling injury. Overmature fruit quickly shows roller bruises, fingerprints, and other signs of handling at harvest. Many growers believe Asian pears are harder to handle than firm peaches and are not suited to large, fast-moving packinghouse lines. Fruit is best packed in the field from picking containers directly into packing boxes or trays. Use boxes with "bubble pads" or paper-covered excelsior pads to prevent rolling in transit.

Some varieties can be stored at 32°F for one to three months without problems. At room temperature (70°F), the fruit begins to soften after 14 to 21 days. Benefits of controlled-atmosphere storage of Asian pears are unknown.

Diseases and pests. Asian pears are susceptible to fireblight and bacterial canker, codling moth, pear psylla, stink bugs, plant bugs, and two-spotted spider mites. When trees are planted too deep in the soil they may die of crown rot. Scab is a problem in Japan, but it is not the same scab species found in California on Bartlett pears and apples.

Rootstocks. There are several rootstocks that can be used in California and the warmer-winter areas of Oregon. All Asian pear varieties will grow on *Pyrus betulaefolia, P. calleryana, P. serotina, P. ussuriensis* and *P. communis* (Bartlett, Old Home × Farmingdale, or Winter Nelis seedling) rootstocks.

Pyrus betulaefolia is preferred for its vigor, large fruit and tolerance of wet soils. In Washington, special cold-hardy *P. betulaefolia* strains are needed. Most Japanese pear varieties are dwarfed about 50 percent on *P. communis* rootstock, so California growers and nurseries prefer *P. betulaefolia* because they like vigorous trees that size fruit easily. Chinese Asian pear varieties like Ya Li are compatible and grow well on either *P. communis* or *P. betulaefolia* rootstock.

Varieties. Asian pear varieties are numerous, with more than 25 known in California and hundreds of varieties known in Asia. Here is a listing of some important varieties available in California, listed from early to late ripening date. Ripening dates given are for Davis, California. In Fresno, ripening will be 7 days earlier; in Oregon and Washington, about 21 to 30 days later.

Ichiban Nashi: An early-maturing, large, brown fruit ripening in mid-July ahead of Shinseiki, Shinsui, and Kosui.

Shinsui: An early-maturing, brown fruit with reasonable size, ripening in mid-July after Ichiban Nashi and before Shinseiki.

Kosui: A small, flat, bronze-russet, early-maturing sweet fruit with tender skin that ripens in mid-July. A strong-growing tree with leaves sensitive to two-spotted spider mites and many sprays.

Shinseiki: A round, yellow-skinned, firm fruit that is early maturing (late July) and stores well up to three months. It looks like 20th Century but is less flavorful. Trees are self-fruitful in the San Joaquin Valley.

Hosui: A very large, juicy, sweet, low acid, bronze-skinned pear that ripens in early August. The tree is extremely vigorous on *P. betulaefolia* and has a wild, loose growth habit. This is a very popular variety in Japan and in California. It is usually very susceptible to fireblight and stores for four to six weeks.

Kikusui: A flat, yellow-green, medium-sized fruit with excellent flavor but a reputation for having tender skin. It ripens in mid-August but fruit has preharvest drop problems.

Yoinashi: A large, brown-skinned fruit with excellent flavor. It ripens in mid-August with 20th Century but sizes much better.

20th Century (Nijisseki): The best-flavored and most popular Asian pear in Japan and California. It is round, yellow-skinned, and easily bruised, but it stores well up to six months. The fruit is more difficult to size than other varieties. It should not be grown on *P. communis* rootstock because it is badly dwarfed. The fruit ripens in mid-August. It grows well on *P. betulaefolia, P. calleryana,* and *P. serotina*. Old trees need spur removal and rejuvenating pruning to maintain fruit size.

Chojuro: An old, firm, brown- to orange-skinned, flat-shaped, highly productive variety that is losing popularity because it is not as juicy as many newer varieties. It matures in mid-August, and bruises easily but will store for five months. Unless the fruit is picked when first yellow-brown in color, it is subject to severe bruising and skin discoloration.

Shinko: The fruit is large and round to slightly flattened with a beautiful bronze-russet skin. Fruit flavor is excellent in hot climates but the fruit stores poorly. The tree is well shaped and extremely productive. It matures during the first week of September, and appears to be nearly resistant to fireblight.

Niitaka: A very large, firm brown-russet fruit. It is noted for its large size, average flavor, and high production. The tree is dwarfed severely on *P. communis* and vigorous on *P. betulaefolia*. Fruit ripens in early September and stores two months. The flowers are pollen-sterile but it sets well when cross-pollinated with most varieties.

Ya Li: A popular Chinese variety. Ya Li is pear-shaped, has green skin, and is quite tender to bruising. It is an early blooming variety that needs cross-pollination from other early flowering varieties like Tsu Li and Seuri. The flavor is sweet and milder than that of other varieties. This is the most important pear variety in China. The fruit ripens in late August and early September, and the harvested fruit stores well.

Tsu Li: A large, football-shaped, green fruit of only fair quality. It has long storage life (6 to 10 months) and gets better the longer it is stored. The fruit ripens in early to mid-September and develops a greasy feeling on the skin. It must be pollinated by Ya Li.

Dasui Li and Shin Li: New patented UC hybrids, very large fruit, greenish to yellow in color. They ripen in late September and early October and store well at 32°F. Trees are extremely vigorous and pollinate each other. For good crops, limited pruning is essential. They grow well on *P. betulaefolia* or *P. communis* roots.

Okusankichi: An old variety from Korea and Japan that ripens in October and stores well. The fruit is brown-russet, somewhat elongated, and slightly irregular in shape. At harvest it has only fair flavor, but flavor improves in storage.

Sources

Planting stock

NOTE: Several commercial nurseries sell Asian pear trees, though some will not sell the trees in small lots.

More information

Berkeley, B. 1985. *Asian pears.* Fowler Nursery, Newscastle, CA.

Beutel, J. A. 1985. *Asian pears.* Washington State Hort. Proc., Wenatchee, WA.

Beutel, J. A. 1988. *Asian pears.* Pomology Dept. Publication, University of California, Davis, CA.

Federal-State Market News Service. 1990. *San Francisco Fresh Fruit and Vegetable Wholesale Market Prices 1990.* California Department of Food and Agri-culture Bureau of Market News and USDA Marketing Service.

Griggs, W., and B. Iwakiri. 1987. *Asian pear varieties in California.* Publication 4068, UC Division of Agriculture and Natural Resources, Oakland, CA.

Sunset Magazine. 1984. The new crunch pears. *Sunset* 84:72–75.

Van der Zwet, T., and N. F. Childers. 1982. *The pear from varieties to marketing.* Horticultural Publications, Gainesville, FL.

Prepared by James A. Beutel.

Baby Corn

Zea mays is a member of the Poaceae (grass) family.

Some corn varieties, such as Baby and Baby Asian, have been developed especially for baby corn production. You can also grow regular sweet corn to pick young. Any well-adapted sweet corn cultivar will work, but some buyers prefer cultivars that produce longer ears. You may also want to choose a variety that bears more than one ear per stalk. Silver Queen, Early Extra-Sweet, Early Sunglow, How Sweet It Is, and Kandy Korn are typical varieties used for baby sweet corn. Keep in mind that the Baby and Baby Asian varieties are different from these sweet corn varieties that are simply picked young.

Sweet corn grown for the edible, tender immature ear is an annual, requiring warm, well-prepared soil and full sun exposure.

Market Information

The super-sweet varieties taste sweet, but they are not the same as the Asian varieties. Baby corn must be sold fresh, and should be part of a delivery made every day to a given market.

Culture

Propagation and care. Sow seeds a bit closer than for regular corn, 6 to 8 inches apart in 30- to 40-inch rows. Do not plant too densely for the amount of nitrogen in your soil or you may get some plants without ears. Water requirements are similar to those for regular sweet corn, except that because it is a short-season crop, baby corn won't need as much total water.

Pick baby corn ears about 1 or 2 days after silks emerge. Beyond that point they will not be sufficiently tender. The variety planted and soil fertility determine the length of the ears. It's best to plant varieties that will bear multiple ears.

Pest and weed problems. Baby corn is a relatively pest-free crop. Ear worm is not a problem, since you harvest before damage would be done. For the same reason, smut is only rarely a problem. Manage other pests, such as seed corn maggot, as you would for sweet corn.

Harvested baby sweet corn ears are 3 to 4 inches long. (Photo: Hunter Johnson)

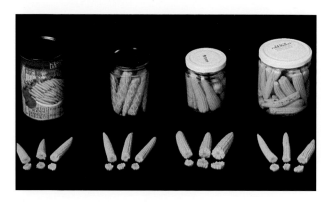

Most baby corn is processed and packaged for later sale. (Photo: Robert Kotch)

Baby corn can be grown fairly easily without chemicals. By incorporating residue soon after you harvest, you can add significant organic matter that will decay much more quickly than if you were to let the plants get old and dry down.

Sources

Seed

Johnny's Selected Seeds, Foss Hill Road, Albion, ME 04910

Le Jardin du Gourmet, P.O. Box 75, St. Johnsbury Center, VT 05863

Nichols Garden Nursery, 1190 North Pacific Highway, Albany, OR 97321

Park Seed Co., Cokesbury Road, Greenwood, SC 29647–0001

Shepherd's Garden Seeds, Shipping Office, 30 Irene Street, Torrington, CT 06790

NOTE: Many seed companies sell regular sweet corn varieties that can be harvested as baby corn.

More information

Sunset Magazine. 1986. What's with these little guys of the vegetable kingdom? *Sunset,* May 1986, pp. 282–84.

*Prepared by Mark Van Horn and Claudia Myers
with assistance from Mike Orzolek, Pennsylvania State University.*

Basil

Ocimum basilicum is a member of the Lamiaceae (mint) family.

Basil, a native to India and Asia, is an annual herb of the mint family. Plants grow in a bushy shape about 18 inches tall, with broad leaves 2 to 3 inches long. Leaves are yellow-green to dark green, depending on the variety and soil fertility. Small white florets produce the plant's dark brown seeds, which self-sow readily. Basil is cultivated for its aromatic leaves, which are used both fresh and dried.

Varieties include *Ocimum americanum* (lemon basil); *O. basilicum* (large-leaf green sweet basil, popular for commercial use); other *O. basilicum* varieties — Anise (anise basil), Cinnamon (cinnamon basil), Crispum (lettuce-leaf basil) Green Ruffles (green ruffles basil), Minimum (bush basil), Nano Compatto Vero (nano compatto vero basil), Piccolo Verde Fino (piccolo verde fino basil), Purple Ruffles (purple ruffles basil), Purpurascens (dark opal basil), and Thyrsiflora (thyrsiflora basil); *O. citriodorum* (lemon basil); *O. gratissimum*; *O. kilimandscharicum* (camphor basil); and *O. sanctum* (holy basil).

Healthy green basil.
(Photo: Gina Good, Harris Moran Seed Co.)

Market Information

The crop is grown commercially worldwide, and there is a market for many types. Basil has medicinal, culinary, aromatic, ornamental, and cosmetic uses. Basil is used as a culinary seasoning and is widely known for its use in Italian cuisine. The cosmetics industry uses basil oil in lotions, perfumes, and soaps. Dried basil can be an ingredient in potpourris and sachets.

For the fresh market, only the highest quality plants (as determined by color and aroma retention) should be used. Basil loses its aroma if the harvested herb is stored too long. Few data are available on herb production or marketing.

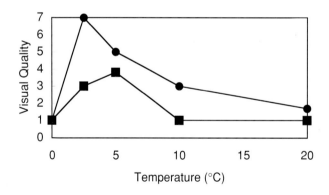

Chart: Effect of holding temperature on the quality of fresh basil after 1 (●) and 2 (■) weeks' storage in perforated polyethylene bags. Visual quality was assessed on a 5-point scale (1, 3, 5, 7, 9, where 1 = low quality and 9 = high quality). From Joyce, Reid, and Katz 1986.

Culture

Climatic requirements. Basil does well in a warm and temperate climate. Cold weather turns the leaves limp and yellow. Basil is extremely sensitive to frost. Growth and yield vary depending on climate conditions, plant type, and cultural and management practices.

Propagation and care. Basil can be direct seeded or transplanted to the field in late spring after all danger of frost has passed. A good, friable, well-tilled and uniform seedbed is required for optimum plant establishment. Seeds should be planted ⅛ to ¼ inch deep. Keep the soil moist to hasten germination. To encourage lateral branching and bushy growth, trim the tops of transplants to 6-inch height before field planting.

Space transplants every 6 inches in rows that are 24 to 36 inches apart. Basil can be grown in high density if mechanical cultivation and seeding are possible. Basil can also be planted 12 inches apart in a bed of 3 rows with 12 inches between rows. Fertilization depends on soil type and previous crop and fertilizer applications. An N–P–K ratio of 1–1–1 is recommended. One approach is to use a broadcast and plowdown application of N–P_2O_5–K_2O at a rate of 120–120–120 pounds per acre, and then to sidedress with 15 to 30 pounds of nitrogen per acre shortly after the first harvest. Jeanine Davis at the Mountain Horticulture Research and Extension Center in North Carolina has worked extensively with commercial basil production. She reports that they have cut back to 75 to 100 pounds preplant and sidedress, or else apply it through the drip system several times during the growing season, 15 to 30 pounds at a time.

High plant populations and mechanical cultivation will help to control weeds. No herbicides are commercially labeled for basil. The soil should be kept moist throughout the growing season in order to avoid low-moisture stress. Drip irrigation is preferred.

Harvest. To ensure a continuous supply of leaves, stagger the field harvests, planting dates, or both.

For dried leaves and the extraction of essential oil, harvest after the flower stalks appear but just before the flowers open. In Mediterranean countries basil is grown as a short-lived perennial and cut 3 to 5 times a year. Harvest foliage above the bottom two to four sets of true leaves. In more temperate regions, basil may be cut only once or twice.

To dry quickly, cut the stalks and spread them on screens out of direct sun. Do not hang them in bunches since the foliage may spoil. Low-temperature, forced-air drying is recommended for leaves and flowering tips to retain color prior to milling or distillation.

The foliage can be cut 4 to 6 inches above the ground to allow for regrowth.

If basil is grown for fresh market on a small scale, cut the stems before flower stalks appear. Loosely bunch the stems, package them, and keep them cool until sold. If the leaves are washed, do not package them with water remaining on the leaves.

Postharvest handling. Basil is extremely susceptible to water loss. For direct marketing, you can maintain quality by standing loosely bunched stems in clean water under ambient temperatures. In most cases, basil must be packaged to reduce water loss. Moist paper lining in the box, vented poly bags, and nonvented fiber cartons are packaging options. Hold at 59° to 68°F for a very short time. Lowering the storage temperature reduces deterioration and water loss.

All basils (including potted plants) are sensitive to damage from low temperatures. Leaves may blacken and lose their aroma. The time required to induce chilling and reduce marketability varies depending on the temperature (see graph). Storage at 50°F (10°C) reduces deterioration without causing much chill damage, and gives a shelf life of 7 to 10 days.

If basil is bulk-packaged in boxes, the heat produced by the respiration of the herb is sufficient to raise the temperature inside the container above that of the storage room. For this reason, boxes containing basil may be kept in storage rooms below 50°F for short periods of time without inducing chill damage. Physical injury at harvest also causes unsightly blackened areas on the sprigs and increases the probability of decay.

Sources

Seed

W. Atlee Burpee & Co., 300 Park Avenue, Warminster, PA 18974

The Cook's Garden, P.O. Box 65, Londonderry, VT 05148

Johnny's Selected Seeds, Foss Hill Road, Albion, ME 04910

Le Jardin du Gourmet, P.O. Box 75, St. Johnsbury Center, VT 05863

Native Seeds, 2509 N. Campbell Avenue #325, Tucson, AZ 85719

Nichols Garden Nursery, 1190 North Pacific Hwy., Albany, OR 93721

Park Seed Company, Cokesbury Road, Greenwood, SC 29647–0001

Seeds Blüm, Idaho City Stage, Boise, ID 83706

Shepherd's Garden Seeds, Shipping Office, 30 Irene Street, Torrington, CT 06790

Sunrise Enterprises, P.O. Box 10058, Elmwood, CT 06110–0058

Taylor's Herb Gardens, 1535 Lone Oak Rd., Vista CA 92084

More information

California Agricultural Statistics Service. 1990. *County Agricultural Commissioner data.* California Dept. of Food and Agriculture, Sacramento, CA.

Economic Research Service, USDA. 1991. *Vegetables and specialties.* TVS-253 and TVS-254, April and August 1991. USDA–ERS, Washington, DC.

Federal-State Market News Service. 1990. *San Francisco fresh fruit and vegetable wholesale market prices 1990.* California Department of Food and Agriculture Bureau of Market News and USDA Marketing Service.

Goldstein, Libby J. 1989. Basil mania. *National Gardening.* Feb. 1989, pp. 30–35, 54–55.

Hampstead, Marilyn. 1984. *The basil book.* Pocket Books, New York, NY.

Herb Market Report. Sweet basil! The perfect annual for the beginning farmer. *The Herb Market Report.* Vol. 2, No. 2, Feb. 1986. Grants Pass, OR.

Kowalchik, Claire, and William Hyltom, et al., eds. 1987. *Rodale's encyclopedia of herbs,* pp. 22–26. Rodale Press, Emmaus, PA.

Newcomb, Duane, and Karen Newcomb. 1989. *The complete vegetable gardener's sourcebook.* Prentice Hall Press, West Nyack, NY.

Simon, J. E. 1984. *A production guide for basil.* VC-002. Horticulture Dept., Purdue University, West Lafayette, IN.

Simon, J. E, and L. E. Craker. 1984. Introduction to sweet basil cultivation. *The herb, spice, and medicinal plant digest.* Vol. 2, No. 2, Fall 1984. Dept. of Plant and Soil Sciences, Univ. of Massachusetts, Amherst, MA.

Simon, James E., and Debra Reiss-Bubenheim. 1988. Field performance of American basil varieties. *The herb, spice, and medicinal plant digest.* Vol. 6, No. 1. Dept. of Plant and Soil Sciences, Univ. of Massachusetts, Amherst, MA.

Prepared by Yvonne Savio and Curt Robinson.

Belgian Endive, French Endive, Witloof Chicory, Chicon

**Cichorum intybus is a member of the Asteraceae (sunflower) family.
Varieties include Daliva, Flash, Flambor, Zoom, and Faro.**

In its first season of growth, this biennial plant produces a rosette of leaves on a thick, fleshy root. At harvest, the roots are lifted from the soil and put in cold storage. They are then "forced" in pits or cold frames. This second, forced growth period produces the edible product, a compact apical bud that if left unharvested would develop into the seed stalk.

Market Information

Current production and yield. The forced crop can be harvested beginning in late September, and harvest can continue through May, peaking from November to March.

Use. Belgian endive is an important vegetable in Europe. It has a delicate, somewhat bitter flavor, and can be cooked in a variety of dishes or used raw in salads. The edible product is the blanched cluster of often yellow-tipped leaves from the compact bud, which is usually called the head or chicon.

Nutrition. Belgian endive is high in fiber, iron, and potassium.

Culture

Climatic requirements. The initial field phase of production benefits from the uniformly cool temperatures of mild days (70°F) and cooler nights (45°F). When forcing, temperatures are critical, with a preference for slightly warmer conditions (60° to 70°F) for the root area than for the tops (55° to 65°F). The temperature around the developing head should be about 5°F lower than it is around the root.

Propagation and care. The production of high-quality, uniform roots requires skill. Growers usually start the crop from seed and grow it somewhat like lettuce. Plants are spaced more closely than for lettuce: every 4 to 5 inches in rows 18 to 20 inches apart, for about 100,000 plants per acre. Growth is like that of romaine lettuce, and proceeds for 3 to 4 months before the roots are dug. Prior to digging, the grower must mow the foliage, taking care to leave the growing point at the apex of the root undamaged. From this tissue the head

Belgian endive shown in cold frames in its second (forced) growth period. (Photo: Vince Rubatzky)

(chicon) will develop. Roots need careful handling during digging and cold storage. Cold storage, whether in the field or in cold storage rooms, is needed to vernalize the tissue. Without proper vernalization, forcing at raised temperatures will be erratic. Maintain a storage temperature of 32°F.

To force, place the roots upright one against the other in soil pits or cold-frame-like structures under a cover of soil. Varieties have been developed that remain compact with no soil cover, so most forcing can be done hydroponically. With hydroponic culture, you can place the roots in trays rather than soil. Both circulating and non-circulating solutions are used to supply moisture and warmth to the roots. After 20 to 30 days the heads are ready to harvest.

Harvest and postharvest practices. The heads are snapped from the roots by hand, and then trimmed and packed. The USDA recommends storage at 36° to 38°F and 95 to 98% relative humidity, with an approximate storage life of 2 to 4 weeks.

Sources

Seed

W. Atlee Burpee & Co., 300 Park Avenue, Warminster, PA 18974

The Cook's Garden, P.O. Box 65, Londonderry, VT 05148

Johnny's Selected Seeds, Foss Hill Road, Albion, ME 04910

Le Jardin du Gourmet, P.O. Box 75, St. Johnsbury Center, VT 05863

Nichols Garden Nursery, 1190 North Pacific Highway, Albany, OR 97321

Nickerson-Zwaan Seed Co., P.O. Box 1787, Gilroy, CA 95021

Royal Sluis, Inc., 1293 Harkins Road, Salinas, CA 93901

Seeds Blüm, Idaho City Stage, Boise, ID 83706

More information

Hanaver, Gary. 1985. European chicory lessons. *Organic Gardening,* 32(5):85–90, May 1985.

Mansour, N. S. 1990. *Witloof.* Vegetable Crops Recommendations. Oregon State University, Corvallis, OR.

Rubatzky, Vincent. 1987. *Witloof: What's that?* Paper presented at the Vegetable Crops Workshop, UC Davis. December 15, 1987.

Rubatzky, Vincent, and Mas Yamaguchi. 1997. *World vegetables, 2nd ed.* Chapman and Hall, New York, NY.

USDA. 1987. *Tropical products transport handbook.* Agric. Handb. 668. USDA, Washington, DC.

Prepared by Vince Rubatzky and Claudia Myers.

Bitter Melon

Momordica charantia **is a member of the Cucurbitaceae (gourd) family,
and a relative of squash, watermelon, muskmelon, and cucumber.
In the United States, varieties are listed as bitter melon, balsam pear, or fu kwa.**

Development of bitter melon varieties has been confined to India and other Asian countries. Many cultivars are available, but little is known about their performance in this country. Cultivars vary in fruit size, shape, and quality, and in yield, maturity, and disease resistance. You should evaluate the varieties available from foreign seed companies and domestic suppliers of oriental vegetable seed to determine which types are best suited to your specific environment.

The plant is a fast-growing trailing or climbing vine with thin stems and tendrils. Male and female flowers are borne separately on the same plant and require insects for pollination. Male flowers appear first and exceed the number of female flowers by about 25 to 1. The flowers open at sunrise and remain open for only one day.

The fruit has a pebbly surface with smooth lengthwise ridges. Immature fruits are light green, oblong, and pointed at the blossom end, and have white flesh. As the fruits begin to mature, their surface color gradually turns to yellow or orange. At maturity, fruits tend to split open, revealing orange flesh and a bright red placenta to which the seeds are attached. Seeds are tan and oval, with a rough etched surface; there are about 150 to 200 seeds per ounce (5 to 7 seeds per gram). The bitter melon probably originated in China or India.

Other names. *Bitter gourd, balsam pear* (United States); *fu kwa* (Chinese); *kerala* (India); *nigai uri* (Japanese); *ampalaya* (Filipino).

Market Information

Current production and yield. Bitter melon is widely grown in China, India, and Southeast Asia. It is grown in small acreages in the United States, primarily in California and Florida. References from India and Southeast Asia report that good yields there are 10 to 12 fruits per plant, or 5 to 7 tons per acre. In Fresno County, California, growers report yields of 15 tons per acre. Total production costs computed for a small farm in Fresno County in 1990 were almost $7,200 per acre, or $4.79 per 20-pound box, assuming 1,500 boxes per acre.

Bitter melon fruit and bunched vine tips. (Photo: Hunter Johnson)

Female bitter melon flowers with developing fruit. (Photo: Hunter Johnson)

Bitter melon at market maturity on the right. Fruit on the left has matured further; its skin has turned yellow, and its interior bright red. (Photo: Hunter Johnson)

Use. In the United States, bitter melon is grown for its immature fruits, which are used in Asian cooking. In other countries, the young leaves are harvested and used as a potherb. The fruit and leaves have a bitter flavor because they contain morodicine, an alkaloid. Alkaloid content can be reduced somewhat by parboiling or soaking fruit and leaves in saltwater. Immature fruit is least bitter. Ripe fruits are extremely bitter, and are reported to be toxic to humans and animals.

Nutrition. Immature fruit is a good source of Vitamin C, and also contains Vitamin A, phosphorus, and iron. The tender vine tips are an excellent source of Vitamin A, and a fair source of protein, thiamin, and Vitamin C.

Culture

Climatic requirements. Bitter melon grows well in warm temperatures similar to those preferred for squash or muskmelons. Frost can kill the plants, and cool temperatures will retard development. Bitter melon normally is grown as an annual crop, but can perform as a perennial in mild-winter areas.

Propagation and care. Plantings are commonly established by direct-seeding in the field. Plant seeds about ½ inch deep. In warm soil, seedlings emerge in a week or less. Transplants can also be used, but plants should be grown by a method that does not disturb the root system during planting. Bare-root plants will not survive well.

A trellis is required to support the climbing vine. The trellis should be about 6 feet high, and constructed from stakes 4 to 6 feet apart with a system of vertical strings running between top and bottom horizontal wires. Rows should be spaced 48 to 60 inches apart, with plants 18 to 24 inches apart in the row. Plants are ideally suited to culture along fence lines of 6- to 8-inch wire mesh.

A deep, well-drained sandy loam or silt loam is the ideal soil, but bitter melon can be grown in any good agricultural soil with proper management. Bitter melon requires the same irrigation practices as squash, cucumbers, or muskmelons, maintaining moisture in the soil's upper 18 inches.

In most California soils, plants require only nitrogen and phosphorus fertilization: 150 pounds of nitrogen and 50 pounds of phosphorus per acre should be adequate. The phosphorus and 50 pounds of the nitrogen should be applied before planting, either by broadcasting and tilling in or by applying in a band a few inches deep and to the side of the plant row. The balance of the nitrogen can be applied in two or more side-dressings with furrow irrigation, or in weekly increments by trickle irrigation. Soil analysis can determine the need for other elements.

Pests and diseases. Bitter melon is susceptible to many of the same diseases and insect pests that affect squash, cucumbers, and muskmelons. It is a host for watermelon mosaic virus, and may be susceptible to other cucurbit viruses. Powdery mildew can be controlled with sulfur dust. The fruits are subject to attack by various fruit flies and fruit rots.

Harvest and postharvest practices. Young fruits should be harvested 8 to 10 days after flowers open, while the fruits are still firm and light green. The fruits will be 4 to 6 inches (10 to 15 cm) long. Beyond this stage, fruits become spongy and more bitter and they lose their market value. The development of mature fruits on the plants may reduce setting of new fruits, so harvesting should be frequent enough to remove fruits at the proper market stage.

The USDA recommends storage at 53° to 55°F at 85 to 90% relative humidity, with an approximate storage life of 2 to 3 weeks. The fruits should be handled and packaged with care, and should be isolated from fruits that produce large amounts of ethylene to prevent postharvest ripening.

Bitter melon is chilling sensitive and should not be held below 50°F. Chilling symptoms include pitting, discolored areas, and a high incidence of decay after the low-temperature storage. Fruits stored at temperatures greater than 55°F tend to continue development and begin to ripen (turn yellow and split open). Controlled atmosphere storage does not provide much benefit toward maintaining green color and overall postharvest quality.

Sources

Seed

W. Atlee Burpee & Co., 300 Park Avenue, Warminster, PA 18974

Seeds Blüm, Idaho City Stage, Boise, ID 83706

Sunrise Enterprises, P.O. Box 10058, Elmwood, CT 06110-0058

Tsang and Ma, P.O. Box 5644, Redwood City, CA 94063

More information

Adlerz, W. C. 1972. *Momordica charantia* as a source of watermelon mosaic virus I for cucurbit crops in Palm Beach County, Florida. *Pl. Dis. Rep.* 56(7):563–64.

Agrawal, J. S., A. N. Khanna, and S. P. Singh. 1957. Studies of floral biology and breeding of *Momordica charantia*. *J. Indian Hort.* 14(1):42–46.

Dhary, A. J. 1971. Midget kerala, the pride of Sorath. *Am. Hort. Mag.* 50(1):46.

Federal-State Market News Service. 1990. *San Francisco Fresh Fruit and Vegetable Wholesale Market Prices 1990.* California Department of Food and Agriculture Bureau of Market News and USDA Marketing Service.

Kays, S. J., and M. J. Hayes. 1978. Induction of ripening in the fruits of *Momordica charantia* L. by ethylene. *Trop. Agric.* 55: 167–72.

Miller, Carey D., Lucille Louis, and Kisaka Yanazawa. 1946. Bitter melon. In *Foods used by Filipinos in Hawaii.* Agric. Exp. Stn. Bull. 98. Univ. of Hawaii, Honolulu, HI.

Morton, J. F. 1967. The balsam pear—An edible, medicinal, and toxic plant. *Econ. Bot.* 21:57–68.

Mote, U. N. 1976. Phytotoxicity of modern insecticides to cucurbits. J. *Maharashtra Agric. Univ.* 1(1):39–42.

Pillai, O. A. A., I. Irulappan, and R. Jayapal. 1978. Studies on the floral biology of bitter gourd (*Momordica charantia* L.) varieties. *Madras Agric. J.* 65(3):168–71.

Rodriguez, D. B., et al. 1976. Carotenoid pigment changes in *Momordica charantia* fruits. *Ann. Bot.* 40:615–24.

Rubatzky, Vincent, and Mas Yamaguchi. 1997. *World vegetables, 2nd ed.* Chapman and Hall, New York, NY.

Sadhu, M. K., and P. C. Das. 1978. Effect of ethral (Ethephon) on the growth, flowering, and fruiting of three cucurbits. *J. HortSci.* 53(1):1–4.

Sundarajan, S., and C. R. Muthukrishnan. 1981. The high-yielding co. 1 bitter gourd." *Indian Hort.* April/June 1981, 25–26.

USDA. 1987. *Tropical products transport handbook.* Agric. Handb. 668. USDA, Washington, DC.

Walters, T. W. 1989. Historical overview on domesticated plants in China, with emphasis on the Cucurbitaceae. *Econ. Bot.* 43(3):297–313.

Zong, Ru-Jing, M. Cantwell, et al. 1990. *Postharvest studies on bitter melon, fuzzy melon, angled luffa, and yard-long bean.* Progress Report, July 1990, Department of Vegetable Crops, UC Davis.

Prepared by Hunter Johnson, Jr., with revisions by Claudia Myers.

Bok Choy

Brassica rapa **Chinensis group is a member of the Brassicaceae (mustard) family.
Varieties include Canton Pak Choy, Pai Tsai White Stalk, Shanghai, Lei Choy, and Pak Choy.**

Bok choy is a non-heading type of cabbage. Looking more like a white swiss chard than a cabbage, bok choy typically has very green leaves with succulent white midribs starting from a bulbous base. The Shanghai variety, commonly called baby bok choy, has green midribs and leaf bases. It probably is native to China and eastern Asia.

Market Information

There is a good market for young or baby bok choy. Baby bok choy is light green in color. It is a staple in the Chinese community. To reach retail stores outside the Chinese community, you need to pack it in smaller boxes (say, 10-pound boxes) rather than the standard WGA pack used in the Chinese groceries.

Current production and yield. California leads the United States in total production of oriental vegetables. The major growing areas are cool-season vegetable production areas such as the Salinas and Santa Maria valleys and the Oceano–Arroyo Grande district. In California, bok choy is produced year round.

Use. Both the dark green leaves and the white ribs are eaten. Bok choy can be stir-fried or steamed, or added to soups or other dishes.

Nutrition. Bok Choy contains 3,000 IU of Vitamin A, 45 mg of Vitamin C, 1.5 g protein, and 105 mg of calcium in each 100 gram raw portion (about 1½ cups chopped).

Culture

Climatic requirements. Average temperatures of 55° to 70°F are favorable. Temperatures exceeding 75°F may cause some leaf tips to burn, and prolonged temperatures below 55°F can cause premature bolting. Bok choy flowering is also sensitive to photoperiods: long days (16-hour days for a month) induce flowering in some cultivars, but short days with warm temperatures keep the plant in the vegetative phase.

Propagation and care. Bok choy is a cool-season crop and can be grown in the same regions and by the same methods as other cole crops, particularly

*Baby bok choy growing in the field.
(Photo: Charlotte Glenn)*

broccoli, cauliflower, and cabbage. Bok choy seeds are direct seeded ½ inch deep and 3 to 4 inches apart in rows 24 inches apart. Plants should later be thinned to 8 or 12 inches apart. Bok choy can be grown from transplants, but special care must be taken to minimize shock in order to prevent premature bolting. The crop requires a rich, well-drained, moist soil with a pH of 5.5 to 7.0. Like all cabbages, bok choy should be encouraged to grow briskly. This is best accomplished by supplying sufficient moisture and nitrogen. Apply at least 1 inch of water weekly.

Fertilization equivalent to 50 to 100 pounds of nitrogen per acre, 50 to 75 pounds per acre phosphorus, and 50 to 75 pounds per acre potassium is recommended. About half the nitrogen should be applied at seeding and the other half shortly after thinning. During cool and wet weather phosphorus deficiencies will occur in bok choy, particularly in heavier soils. All of the phosphorus and potassium should be applied before or at the time of planting.

Bok choy matures 40 to 60 days after sowing. At harvest, take care to avoid excessive cracking of the ribs. To harvest, cut the entire head of the plant off at ground level. Bok choy heads should be harvested before the first hard frost. After harvesting, trim the heads and remove all damaged or unsightly leaves and root remains.

Harvest and postharvest practices. The USDA storage recommendation is 32°F at 95 to 100% relative humidity, with an approximate storage life of 3 weeks. The crop is commonly shipped in wooden crates, 18 to 24 heads per container, and wrapped in parafilm paper. Each container weighs 50 to 60 pounds. The containers need to be cooled properly to prevent produce decay.

Sources

Seed

W. Atlee Burpee & Co., 300 Park Avenue, Warminster, PA 18974

The Cook's Garden, P.O. Box 65, Londonderry, VT 05148

Johnny's Selected Seeds, Foss Hill Rd., Albion, ME 04910

Native Seeds, 2509 N. Campbell Avenue #325, Tucson, AZ 85719

Park Seed Co., Cokesbury Road, Greenwood, SC 29647–0001

Seeds Blüm, Idaho City Stage, Boise, ID 83706

Shepherd's Garden Seeds, Shipping Office, 30 Irene Street, Torrington, CT 06790

Sunrise Enterprises, P.O. Box 10058, Elmwood, CT 06110–0058

Tsang and Ma, P.O. Box 5644, Redwood City, CA 94063

More information

Federal-State Market News Service. 1988, 1989. *Los Angeles fresh fruit and vegetable wholesale market prices 1988 and 1989.* California Department of Food and Agriculture Bureau of Market News and USDA Marketing Service.

Federal-State Market News Service. 1987, 1988. *San Francisco fresh fruit and vegetable wholesale market prices 1987 and 1988.* California Department of Food and Agriculture Bureau of Market News and USDA Marketing Service.

The Packer. 1990. *1990 Produce availability and merchandising guide.* Vance Publishing Corp., Overland Park, KS.

Rubatzky, Vincent, and Mas Yamaguchi. 1997. *World vegetables, 2nd ed.* Chapman and Hall, New York, NY.

Stephens, James. 1988. *Minor vegetables.* Cooperative Extension Bulletin SP-40. University of Florida, Gainesville, FL.

USDA. 1987. *Tropical products transport handbook.* Agric. Handb. 668. USDA, Washington, DC.

Prepared by Claudia Myers, Imo Fu, and Louie Valenzuela.

Bottle Gourd, Calabash Gourd, Cucuzzi

Lagenaria siceraria is a member of the Cucurbitaceae (gourd) family.

The bottle gourd probably originated in south-central Africa. It has been used for thousands of years by people in the topics and subtropics. The bottle gourd is an annual climbing vine with large leaves (6 to 12 inches diameter) and a lush appearance. The foliage is covered with soft hairs. The branched vines and tendrils spread or climb, growing from 10 to 50 feet long. The large, white flowers characteristic of the plant open in the evening and remain open until the following midday. Flowers are pollinated by insects. The background color of the gourds is either light or dark green. The fruit size varies from 2 to 12 inches in diameter and from 4 to 36 inches in length. The fruit can have a sterile (seedless) neck that varies from a few inches up to 15 inches long and from 1 to 2 inches wide. Wider necks usually contain seeds. Gourds may be round, bottle-shaped, dumbbell-shaped, crooknecked, coiled, or spoon-shaped.

Other names. *Yugao* (Japanese); *po gua* (Cantonese Chinese); *upo* (Filipino); *bau* (Vietnamese).

Market Information

The young fruit tastes just like zucchini. Although most people don't know what to do with it, young bottle gourd has had a steady market over the years.

Use. Young bottle gourd fruit is eaten as a boiled vegetable. When bottle gourds are to be used as containers, they may be constricted with bands to make a particular shape. A gourd can be preserved with a coat of wax, lacquer, or shellac. Well-treated gourds can become durable containers, churns, ladles, spoons, pipes, carved objects, or musical instruments.

Nutrition. The bottle gourd is not an important source of any nutrient. The composition of 100 grams of young bottle gourd includes 20 mg calcium and 6 mg of Vitamin C.

Culture

Climatic requirements. The bottle gourd can be grown anywhere in California during frost-free periods of the year. For best growth, the gourd needs hot days and warm nights. Under good conditions it will show vigorous and rapid growth.

Bottle gourd growing on a trellis in Taiwan. (Photo: Imo Fu)

Bottle gourd growing on the ground in California. Note the varying sizes and shapes. (Photo: Hunter Johnson)

Propagation and care. Space plants 9 feet apart. Plant seeds 1½ inches deep in raised beds or mounds in full sun. In short-season areas, start seed indoors in pots 3 to 4 weeks before average last frost. A trellis is advisable, but vines may be allowed to run on the ground. Soil, water, and fertilizer needs are similar to those of squashes, melons, and cucumbers.

Fruit size is related to plant vigor, the number of fruits set, and the time of year. To obtain larger fruits, reduce competition by pruning off other newly set young fruits.

With ground culture the use of mulch helps to keep the fruit from rotting, but fruit often form away from the mulch. Harvest gourds when the vines are dry. Cut some stem with each gourd and hang the gourd to dry in a cool, airy place, such as a covered patio or beneath trees. Depending on temperatures, the gourds may be adequately dried in 2 to 4 weeks.

Sources

Seed

W. Atlee Burpee & Co., 300 Park Avenue, Warminster, PA 18974

Native Seeds, 2509 N. Campbell Avenue #325, Tucson, AZ 85719

Sunrise Enterprises, P.O. Box 10058, Elmwood, CT 06110–0058

Tsang and Ma, P.O. Box 5644, Redwood City, CA 94063

More information

Gore, Willma Willis. 1984. Out of her gourd? *California Farmer* 267(6):8ff.

Martin, Franklin W. 1979. *Vegetables for the hot, humid tropics. Part 4. Sponge and bottle gourds,* Luffa *and* Lagenaria. Agricultural Research (Southern Region), Science and Education Administration, USDA, New Orleans, LA.

Stephens, James. 1988. *Minor vegetables.* Cooperative Extension Bulletin SP-40. University of Florida, Gainesville, FL.

Prepared by Claudia Myers and Ed Perry.

Canola

Brassica napus (Argentine type) and *Brassica campestris* (Polish type) are members of the Brassicaceae (mustard) family. Varieties are suited for planting at specific times of year. Those that have been grown in California include Legend, Moneta, and Westar.

Canola is edible oilseed rape. It contains small, nontoxic amounts of the same erucic acid and glucosinolates that make conventional rapeseed toxic to humans and animals. Canola resembles a turnip plant in the fall, but without the large root. When mature it is 4 to 6 feet tall, with tan stems and seed pods that are 2 inches long and contain 15 to 20 seeds each. Canola flowers are mustard-yellow.

Market Information

Imports supply most of the U.S. market for canola, which reached 300 million pounds in 1988. Canada produces 15 percent of the world's canola. The European Economic Community produces 17 percent. Domestic production, centered in Minnesota and North Dakota, accounts for 1 percent of the world market. Some acreage is planted to canola in the Pacific Northwest. There is a large potential market in Japan for June-harvested canola such as is grown in California.

California Oils Corporation in Richmond is the only canola processor in California. In 1990 there were three others in the United States and Canada: Canola Processors (Central Soya) in Chattanooga, Tennessee; ADM in Windsor, Ontario; and ADM in Velva, North Dakota. Growers will need a local elevator that will ship to the processor.

Current yield and production. Soils that produce the highest yields for wheat will do the same for canola. Most production is for cash sales at the market. Processors offer forward contracts to growers. All pricing relates to the Winnipeg Commodity Exchange in Manitoba, and is listed daily in the Wall Street Journal.

Use. In California, rapeseed and mustards have long been used as cover crops or green manure crops for orchards and vineyards. When its high oil content is removed, canola is a high-quality, highly palatable feed concentrate of 37 percent protein. Canola oil is a superior cooking oil. It has a bland flavor, light color, and delicate aroma. Both canola species produce seed high in polyunsaturated fatty acids.

Canola in flower, one day after irrigation. (Photo: Tom Kearney)

Culture

Climatic requirements. Canola is adapted to the cool extremes of the temperate zone. The crop emerges at an optimum temperature of 50°F, with a low emergence temperature of 41°F. Minimum temperature for growth is 30°F. Snow covers are suggested for winter annual varieties, although they have been produced in Michigan without snow covers. When the plant has about 6 leaves and is about 5 inches tall, it has adequate root reserves to survive winter dormancy.

Propagation and care. To avoid contamination, do not plant canola near wild mustard crops that have high seed levels of erucic acid. If you practice summer fallow, plant canola after the fallow, with a cereal crop following the canola crop. A firm, level, weed-free, moist seedbed is ideal. Seedlings cannot compete with weeds that germinate after fall rains. If there are no early fall rains, a dry seedbed is acceptable. Seeding time is similar to that of wheat: early November to early December. Sow as deep as 1 inch at rates of 5 to 10 pounds per acre (*B. napus*) or 4 to 5 pounds per acre (*B. campestris*). Space the rows 6 to 12 inches apart. Apply 25 pounds of nitrogen at planting after summer fallow and 30 pounds per acre as a top dressing. After a cereal crop, double these amounts. For phosphorus, use 50 pounds per acre if broadcast and 25 pounds per acre if supplied with the seed. Fertilizers containing sulfur should be used where legumes respond to sulfur.

Contamination with weedy mustard relatives is a serious problem. The oil and protein quality may be lost, causing crop values to fall. It is important to plant high-quality seed free from outcrossing to weedy mustards and non-canola rapeseed.

Harvest and postharvest practices. Canola dries from bottom to top. During harvest, it requires careful management since it is susceptible to shattering when mature. Most farmers direct-harvest when the moisture content reaches 8 to 10 percent if they are using a standard combine.

In 1990–1991 most of California's commercial acreage was sprayed with Spodnam DC (active ingredient: polymer of cyclohexane, 1-methyl-4-[1-methylethyl]) to prevent shattering. The chemical seals pods to keep them from shattering, and is applied when pods range from green to not completely mature. The 1990–1991 crop was direct-combined. Canola can also be swathed first and then direct-threshed with a combine. You should windrow canola at 35 percent moisture, when the crop is greenish brown with firm seeds.

Canola harvesting requires particular attention to crop moisture levels, combine settings, and environmental conditions at the time of harvest. The harvester's reel speed and forward movement speed should be equal, with the cutter bar set just below the level of the seed pods. Once the crop reaches 35 percent moisture it will mature in 4 to 6 days, so windrowing should not be delayed. Rapeseed is ready to be picked up and threshed when the seed contains 10 to 8 percent moisture. Adjust the harvester's pickup speed and forward travel to eliminate any tearing of the swath. Reduce the combine cylinder speed to two-thirds of that used for cereals. Open the combine's concaves to reduce breakage of the stems and pods. The seeds are lightweight, so fan speeds should be reduced and fan louvers partially closed to shake the seed out of the chaff. The top sieve or chaffer should be slightly open, and the lower sieve nearly closed. Always consider wind speed and direction, and make frequent adjustments to compensate for these factors.

Store harvested canola in tight bins, and inspect it often to prevent heating and spoilage. The seeds are small and not conducive to good air flow in bins, so the wet seeds must be arranged in thin layers to dry. Blowing air through the bins might also be helpful. After harvesting, chop all of the field residue fine and spread it uniformly over the field to prevent any toxic effect on following crops. Soybeans have been found to have better yields when planted after canola.

Pests and diseases. White rust–downy mildew complex, sclerotinia stem rot, and blackleg all affect canola. Some newer canola strains have a higher tolerance to blackleg. Practice crop rotation to prevent disease in the crop, separating canola crops with at least two cereal crops. Good water and fertilizer management also limits the incidence of disease.

Cabbage aphids can hurt canola crops. If you use insecticides, be careful not to harm the bees that will be present when canola is in bloom. Monitor aphids closely after flowering, since restrictions apply to some aphicides as the crop nears harvest. Aphid honeydew from a heavy infestation can affect combine performance and harvesting.

Weed control is crucial. Tillage, establishment of a good stand, and weed control in previous crops are all good weed-prevention practices.

Sources

Seed

Bonis and Company Ltd., Box 217, Lindsay, Ontario, K9V 5Z4, Canada

Kaystar Seed, Division of Kaystar Corp., 702 Third Street SW, Huron, SD 57350

Northern Sales Co. Ltd., 5th Floor–200 Portage Ave., Winnipeg, Manitoba R3C 3X2, Canada

SeedTec International Inc., Woodland, CA 95695

More information

Albright, Letha. 1989. Canola: Promising alternative oil crop. *Missouri Farm*. Missouri Farm Publishing Inc., Clark, MO. 6(4):40–41. July/Aug 1989.

Ameri-Can Pedigreed Seed Co. 1989–1990. *The canola report.* Ameri-Can Pedigreed Seed Co. Memphis, TN. 1(3, 4), 2(1, 2).

Anon. 1990. *MidAmerica Farmer Grower.* SJS Pub. Co., Perryville, MO. July 12, 1990.

Frank, Robert. 1990. Canola market and outlook. *MidAmerica Farmer Grower.* SJS Pub. Co., Perryville, MO. July 12, 1990, p. 16.

Frank, Robert. 1989. Canola showing alternative potential." *Downstate Farmer.* Robert Peach, Columbia, IL. April 1989, p. 12.

Knowles, P. F., and T. E. Kearney. 1979. *Rapeseed production in California.* Agronomy Progress Report No. 102, October 1979. University of California, Davis, CA.

Oplinger, E. S., L. L. Hardman, E. T. Gritton, T. D. Doll, and K. A. Kelling. Canola (rapeseed). *Alternative field crops manual.* Departments of Agronomy and Soil Science, College of Agriculture and Life Sciences and Cooperative Extension Service, University of Wisconsin-Madison, WI.

SeedTec International. *Management/production guide for winter and spring canola.* SeedTec International Inc., Woodland, CA.

NOTE: Other information is available from the United States Canola Association, 1150 Connecticut Avenue NW, Washington, DC 20036.

Prepared by Tonya Nelson.

Caper

Capparis spinosa **is a member of the Caperaceae family. No detailed information is available on varieties. Some capers have spines, other have none; both types appear to produce equally well.**

Capers are native to the Mediterranean area and the tropics. The plant, a deciduous dicot, grows about 2 feet tall and has a very deep root system. Vines are 7 to 10 feet long with shiny green oval leaves. The white or purple flowers resemble rock rose blossoms. The flowers are bisexual and have a lifespan of 24 to 36 hours. Each plant produces hundreds of flowers each season. The mature fruit is 2 to 3 inches long and ½ to ¾ inch in diameter. It starts out green, but turns purple when mature. Each fruit contains 200 to 300 seeds.

Market Information

The smaller the caper bud, the higher its quality and price. The consumer price for 7 ounces of processed, good-quality capers is close to $5.

Caper flowers are insect pollinated. (Photo: D. Kontaxis)

Current production. Capers are produced in Morocco, Spain, and Italy. In Greece, Cyprus, and Turkey the plant is well adapted, but is not cultivated commercially. The United States imports more than $20 million worth of processed capers each year.

Use. Capers are used as a condiment in salads and sauces, or with meat and fish. They are also used in the manufacture of cosmetics and medicines. Some *Capparis* species are poisonous. Depending upon the part of the plant that is used, capers can be considered a vegetable (the edible shoots) or an herb (the processed buds). Because of its attractive flowers and foliage, the caper plant can be used in ornamental plantings. It may be used also to control soil erosion, especially on slopes where irrigation is difficult and soil erosion is more pronounced.

Culture

Propagation and care. The plant needs little care. It is drought resistant but requires good drainage. It has few disease and insect pest problems.

Propagation is best accomplished from roots or cuttings because of the variability in seed-propagated plants. Root the cuttings in a greenhouse for at least one year, and then plant in the field on an 8-

Caper plant. (Photo: D. Kontaxis)

by-8-foot grid during February or March. In the first two summers new plants require two to three irrigations. Older plants need less irrigation except in dry years and very hot summers. Spring fertilization is advisable, with irrigation after each application.

Seedlings are very temperamental when transplanted, and some may die. In order to reduce this loss, transplant with soil attached to the root system, and water immediately after transplanting.

Germination of caper seeds is difficult, but the following methods have resulted in 40 to 75 percent germination. First, heat some water to 110° or 115°F, and put the seeds into the warm water to soak for at least 12 hours, during which time you can allow the water to cool to room temperature. After 12 hours, discard the water, wrap the seeds in a moist towel, place them in a plastic bag, and keep them in a refrigerator for 65 to 70 days. Then take the seeds out of the refrigerator and soak them in warm water (110° to 115°F) overnight.

Plant the seeds about ¼ to ½ inch deep in a soil mix of 50–25–25 parts of UC soil, perlite, and sand, respectively (planting mix can be used instead of UC soil mix). Use 6-inch clay pots or deep flats. Water well and keep in a warm area (70° to 85°F), in partial to full sun. Do not allow the top of the soil to crust over. Keep soil moist. Germination should start within 3 or 4 weeks, and may continue for 2 to 3 months. Not all seeds will germinate at the same time.

Let the seedlings grow to 3 to 5 inches tall before transplanting. If seedlings are too crowded in the clay pot or flat do not pull them—use scissors to cut off the smaller, less-vigorous plants, leaving the root systems of the remaining seedlings undisturbed.

Transplant the seedlings to individual 1-gallon containers, using the same planting mix described above. When transplanting, disturb the root system as little as possible, keeping some original soil around each transplanted seedling. Good soil drainage is essential to prevent root rot. Pack the soil tightly around the transplanted seedling and water immediately. Cover each container with a plastic bag. Keep in a shaded spot in spring or summer or in a warm area (70° to 85°F) in winter. Keep the plastic bag in place for 1 week. At the end of the week, cut off the top of the bag so that the seedling will be exposed gradually to the natural environment. In another 10 days enlarge the opening in the plastic bag. One week later remove the bag entirely, keeping the plant in a shaded area. Keep the plants in their 1-gallon containers and then transplant them in early spring, after the last frost, when soil is workable.

Plant the capers in elevated rows. The rows should be 8 to 10 feet apart, and the plants in each row should be 8 to 10 feet apart within the row. Water frequently, but make sure that drainage is adequate and fertilize two to three times during the spring and summer months. Irrigation is essential for the first 2 years of development.

Do not prune the young plants for the first two years. Prune 3-year-old or older plants during November or December. Cut the canes back, but only to 3 or 4 inches from the crown—cutting the canes all the way to the crown may kill young plants.

Harvest and postharvest. In the spring, pruned plants develop tender new shoots that can be eaten as a vegetable. Pick the buds from mid-May to mid-August. A 2-year-old plant produces a few buds; a 3-year-old plant produces just over 2 pounds in a year; and a plant older than 4 years may produce more than 20 pounds of buds. Unopened buds are picked by hand, sorted, and pickled in brine.

Curing and packing. Sort the harvested caper buds by size: small, medium, or large. Place sorted buds in a strong brine (1¼ to 1½ lb salt per gallon of water—a salt hydrometer or salometer will show 15 to 18 percent or 60 to 72 degrees). Keep the capers in brine for 30 to 45 days. They can stay in the brine for 12 months without damage, but make sure they remain submerged. When you remove them from the brine, rinse them in running water for several minutes to remove excess salt.

Pack capers into small jars up to the shoulders of the jars. Cover the capers with an acidified solution of 1 gallon 5 percent vinegar in 2½ gallons of water and 1½ teaspoonful of salt. Leave ¼ to ½ inch headspace between the top of the liquid and the rim of the jar. Adjust lids and screw them down tight. Do not over-tighten the lids. Jars of capers in acidified brine can be left in the refrigerator for as long as 6 months, or they can be pasteurized by submersion in a hot water bath of 170° to 175°F. Once the jars are in the water bath and the water temperature returns to 170°F, process the jars for 30 minutes. Remove jars from the water bath and allow them to air-cool. The shelf life for pasteurized capers is about 1½ years at room temperature.

Pests and disease. In California, caper plants can be damaged by gray mold fungus, nematodes, and insect pests such as imported cabbageworm, black vine weevil, and flea beetle. Gophers, snails, and slugs can also cause leaf destruction.

Sources

Seed

Park Seed Company, Cokesbury Road, Greenwood, SC 29647–0001

NOTE: Some California nurseries may carry a very limited number of caper plants.

Prepared by Demetrios G. Kontaxis.

Cardoon

**Cynara cardunculus is a member of the Asteraceae (sunflower) family.
Varieties include Plein Blanc Inorme and Italian Dwarf.**

Closely related to the artichoke, cardoon is grown for its edible leaf petioles rather than for the undeveloped flower buds. The thistle-like plant grows vigorously, reaching a height of 3 to 5 feet and spreading to a diameter of 6 feet. It is a perennial plant native to the Mediterranean region, where it is widely grown.

Market Information

The harvested grouping of petioles looks like a large, waxy celery. Cardoon's flavor is similar to that of the artichoke: the cooked flesh is relatively bland, but alkaloids in its tissues can impart a bitter flavor.

Use. Young tender leaves, leaf petioles, and undeveloped tender flower stalks are edible. Grown in warm weather, these tissues can have a strong flavor. Cooking removes the bitter flavor (the alkaloids are destroyed by heating). A popular way to prepare cardoon is to steam or boil the young small leaves and tender stalks together. If you boil them, discard the water once it has drawn off the bitter flavor, and finish the boiling in a fresh pot of water. Cardoon can be used to flavor soups, chilled after cooking and served with a vinaigrette dressing, or prepared hot, topped with a cream sauce. Cardoon can also be marinated and eaten uncooked in salads. As "cardoni," it is a delicacy in some parts of Italy.

Nutrition. Cardoon contains 120 IU of Vitamin A, 2 mg of Vitamin C, 0.7 mg of protein, and 70 mg of calcium in a 100 g raw, edible portion (about ½ cup shredded).

Culture

Climatic requirements. Botanists do not consider cardoon to be a cool-season crop, but it is cultivated as if it were. Cardoon will develop too much alkaloid and become inedible when grown in hot summertime conditions. The crop should be planted so it will develop and be harvested under moderately cool conditions. Cool temperatures ranging from 55° to 65°F are preferred. Freezing temperatures below 28°F will damage or kill aboveground parts of tender, non-acclimatized cardoon.

*Cardoon grows to a height of 3 to 5 feet.
(Photo: Hunter Johnson)*

Propagation and care. Cardoon is commonly propagated from stem portions having axial buds or from suckers, which are rooted offshoots that develop on old stems and are then removed. Plants can also be produced from seed. Sow the seed indoors and transplant seedlings to the field when they are 3 to 4 inches tall. Place either transplants, seedlings, suckers, or stem portions 20 inches apart in rows 3 to 4 feet wide. If you are planting seed directly into the field, sow several seeds in clusters at 20-inch intervals in the row. Once the seedlings are established, thin each cluster to a single plant. With approximately 700 seeds per ounce, planting an acre of cardoon requires 4 to 5 pounds of seed.

Do not harvest cardoon in warm weather: the flavor will be bad. Suitable harvest dates are the same as those for artichoke. Like artichoke, cardoon is frost-sensitive.

Cardoon does best with rich, well-drained soil, plenty of water, and room to grow. Plants may require 6 months to reach full maturity. Reduce weed competition by weeding regularly. Two or three weeks before harvest, when the plants are approaching maturity, you can blanch the stalks to give the harvested produce a milder flavor. To blanch, tie the outer branches together about a foot from the top of the plant and wrap the base of the stalks with heavy paper, burlap, or other material to a height of 12 to 18 inches. Avoid covering the leaf tips. To harvest, cut the plants off below the crown and peel away the coarse outside leaves. The product will resemble celery, and can be trimmed to a length of 12 to 16 inches. About 1 foot of tender stalk is usable.

Sources

Seed

Epicure Seeds, Ltd., P.O. Box 450, Brewster, NY 10509

Exotica Seed Co., 8033 Sunset Boulevard, Suite 125, West Hollywood, CA 90046

Le Jardin du Gourmet, P.O. Box 75, St. Johnsbury Center, VT 05863

Nichols Garden Nursery, 1190 North Pacific Highway, Albany, OR 97321

Seeds Blüm, Idaho City Stage, Boise, ID 83706

More information

Glenn, Charlotte, and Georgeanne Brennan. 1988. *Le marché seeds international spring '88 catalog.* Dixon, CA.

Mansour, N. S. 1990. *Cardoon.* Vegetable Crops Recommendations, Oregon State University, Corvallis, OR.

Rubatzky, Vincent. 1983. Grow cardoon for piquant flavor and beauty. University of California news release, June 24, 1983.

Rubatzky, Vincent, and Mas Yamaguchi. 1997. *World vegetables, 2d ed.* Chapman and Hall, New York, NY.

Stephens, James. *Minor vegetables.* 1988. Cooperative Extension Bulletin SP-40, University of Florida, Gainesville, FL.

Prepared by Vince Rubatzky and Claudia Myers.

Celtuce, Asparagus Lettuce

***Lactuca sativa* var. *augustana* is a member of the Asteraceae (sunflower) family.**

Celtuce looks like a cross between celery and lettuce. The outer leaves resemble those of loose-leaved lettuce, but are a lighter green.

Market Information

Use. Young, tender celtuce leaves can be eaten in salads. The soft, translucent green central core can be eaten fresh, sliced or diced in a salad. In China, the fleshy stem is cut into sections and boiled or stewed. Cooked, celtuce tastes like a cross between a mild summer squash and an artichoke.

The Packer's 1990 Produce Availability and Merchandising Guide listed only one shipper of celtuce in the United States: Plantation Spice Growers Inc. of Goulds, Florida.

Culture

Climatic requirements. Celtuce is a cool-season crop. It requires a mean temperature of 50° to 60°F.

Propagation and care. Plant celtuce from seed in the fall, winter, or early spring. Space the seeds about 8 inches apart in the row, and treat the new crop like regular lettuce. Soon after the outer leaves develop, a central stalk bearing tiny leaves at the top starts to grow. When the stem is 12 to 18 inches long, cut it down into the leafy part of the plant.

Pick the celtuce stem when it reaches a length of 12 to 18 inches. (Photo: Hunter Johnson)

Peel the stem, since its outer edges contain a bitter, milky sap. The soft, translucent, green central core is the edible part. The leaves may be eaten in salads or as greens when they are young and tender, but as they mature they grow bitter and become inedible.

Sources

Seed

W. Atlee Burpee & Co., 300 Park Avenue, Warminster, PA 18974

Nichols Garden Nursery, 1190 North Pacific Highway, Albany, OR 97321

Seeds Blüm, Idaho City Stage, Boise, ID 83706

Sunrise Enterprises, P.O. Box 10058, Elmwood, CT 06110–0058

Tsang and Ma, P.O. Box 5644, Redwood City, CA 94063

More information

Herklots, G. A. C. 1972. *Vegetables in South-East Asia.* George Allen and Unwin Ltd., London.

The Packer. 1990. *1990 Produce availability and merchandising guide.* Vance Publishing Corp., Overland Park, KS.

Rubatzky, Vincent, and Mas Yamaguchi. 1997. *World vegetables, 2d ed.* Chapman and Hall, New York, NY.

Stephens, James. *Minor vegetables.* 1988. Cooperative Extension Bulletin SP-40, University of Florida, Gainesville, FL.

Prepared by Claudia Myers (adapted from James Stephens's Minor Vegetables*).*

Chayote, Mirliton, Vegetable Pear

Sechium edule **is a member of the Cucurbitaceae (gourd) family.**

The chayote plant has climbing vines and leaves that resemble those of cucumber. The plant is a perennial in the tropics, where stems have tendrils and can grow 50 feet long. The plant produces separate male and female flowers that are pollinated by bees. The light green pear-shaped fruit contains a single edible seed about 1 or 2 inches long. Fruits of different varieties range from almost smooth to deeply ridged, and from cream-colored to green, and may or may not be covered with nonsticking prickles. The mature fruit is 3 to 8 inches long, weighing from 8 ounces to more than 1 pound. Chayote was cultivated centuries ago by the Aztecs and Mayans of Central America.

Other names. *Tao tah* (Hmong); *hayato uri* (Japanese); *fut shau kua, ngow-lai choi, tsai hsio li* (Chinese); *sayote* (Filipino); *xu-xu, trai su* (Vietnamese); *cho cho* (West Indies); *mirliton* (Louisiana).

Market Information

Chayote is found in many ethnic markets, including Asian, Indian, Caribbean, and Latin American markets. It is also a familiar sight in major supermarkets. The prickly varieties are seldom grown, but the unusual appearance and varied flavors of their fruit could make them a good crop for small growers.

Current production and yield. Chayote is available year-round, but is most abundant from September through May. California, Florida, Costa Rica, and Mexico are the main suppliers. Quantities of fresh and frozen chayote imported into the United States have increased steadily, from 5,232,000 pounds in 1980 to 13,543,000 pounds in 1988. Chayote yields 15 to 20 tons of fruit per acre.

Use. Chayote fruit has crisp, pale flesh and a flavor that is a blend of apple and cucumber. In many countries the young shoots, flowers, seeds, and roots are eaten. Chayote is served in many ways: creamed, fried, baked, and pickled, and in salads, soups, stews, and pies.

Chayote on sale at a grocery store. (Photo: Charlotte Glenn)

Nutrition. Chayote is low in calories: only 40 calories per cup. It is also low in sodium and is a good source of fiber. Its nutritional value is similar to that of summer squash.

Culture

Climatic requirements. Chayote is a warm-season crop. Less-favorable climatic conditions promote luxuriant vine growth but much lower fruit production. Plant chayote in early spring or when the ground is sufficiently warm. The fruit may need to be covered completely to prevent cold damage.

Chayote blooms in the shorter daylight of late summer and autumn days, but the fruit must mature before winter. In the San Diego and Los Angeles areas, the plants bloom in late August or September. Winter frost causes the vines to die back, but they return with warmer temperatures. The same plant can continue to produce fruit for several years. In the Panama Canal Zone, where day length is about 12.5 hours year-round, chayote blooms and produces fruit every month of the year. The fruit reaches full size about 30 days after pollination.

Louis Aung et al. (see *More information*) write that chayote may be grown in temperate climates under a regime of artificially controlled day lengths. After 6 to 8 weeks of growth, the vines are shaded with dark cloth on a frame to limit sunlight to 8 hours a day for 4 to 6 weeks. Once flowers develop, the shades can be removed and vines can grow in normal daylight.

Propagation and care. Trellises or structures similar to grape arbors are required for chayote production. Vines are trained over the top of the trellis, and fruits are harvested from below. Some growers use vertical trellises instead. Plant one fruit per hill, with hills 12 feet apart and 12 feet between rows. Plant the whole fruit on its side with the stem end sloping upward. Fruits obtained from a supermarket will sprout when kept in subdued light, and are ideal for planting. For greater uniformity of plant type, you may want to plant from stem cuttings instead.

Fertilization requirements for chayote are similar to those for summer squash, with both nitrogen and phosphorus applications needed in many growing areas. Depending on temperature and soil texture, irrigation may be required once or twice a week. Overmature chayote fruits will sprout on the vine but are still edible if properly prepared.

Pests. Leaf-eating beetles and snails occasionally reduce plant growth. Leaf-eating insects seldom require control. Nematodes occasionally reduce chayote yields and should be controlled several weeks before planting.

Harvest and postharvest handling. Ripe fruits are selected by size and pulled from the vines. Harvested chayote fruits can be wrapped individually in tissue paper or poly bags to reduce friction and water loss and placed in single-layer flats. Fruits stored in poly bags may decay since the bags increase moisture condensation. Seed germination is a storage problem if the fruit is held at temperatures above 56° or 58°F. Chayote fruit is also susceptible to chilling injury. Surface bronzing affects fruits held at 36° to 41°F, and surface pitting, decay, and internal browning appear in fruits held at 41° to 45°F. With a storage temperature of 45° to 50°F, you can expect chayote to have a shelf life of up to 4 weeks.

Sources

Seed
NOTE: Plant whole fresh fruit or propagate from stem cuttings.

More information
Aung, Louis H., Amelia Ball, and Mosbah Kushad. 1990. Developmental and nutritional aspects of chayote. *Economic Botany* 44(2): 157–64.

Cantwell, Marita. 1987. *Postharvest handling of specialty crops: chayote.* Perishables Handling No. 61. Vegetable Crops Dept., UC Davis.

Economic Research Service. 1989. *Vegetables and specialties: Situation and outlook yearbook.* USDA Economic Research Service, Washington, DC.

Hall, B., and J. MacGillivray. 1958. *Chayote production in California.* UC Agricultural Extension Service, OSA #36.

The Packer. 1989. *1989 Produce availability and merchandising guide.* Vance Publishing Corp., Overland Park, KS.

Rubatzky, Vincent, and Mas Yamaguchi. 1997. *World vegetables, 2d ed.* Chapman and Hall, New York, NY.

Stephens, James. *Minor vegetables.* 1988. Cooperative Extension Bulletin SP-40, University of Florida, Gainesville, FL.

USDA. 1987. *Tropical products transport handbook.* Agric. Handb. 668. USDA, Washington, DC.

Prepared by Claudia Myers and Keith Mayberry as an update of Bernarr Hall and John MacGillivray's Chayote Production in California.

Chinese Broccoli, Kailan, Gai-lohn, Chinese Kale

Brassica oleracea Alboglabra group is a member of the Brassicaceae (mustard) family. Varieties are usually listed as Gai-lon (or Gai-lohn), Kailan, Chinese Broccoli, or Chinese Kale.

The Chinese broccoli plant resembles our more familiar broccoli, except that the leaves are a bit broader, the stems are longer, and the head is much smaller. Flowers form first in diminutive heads, and then elongate rapidly into stalks bearing yellow or white flowers.

Other name: *Pak kah nah* (Hmong).

Market Information

Chinese broccoli is not well known in the mainstream market but is very common in Asian community markets, where its quality may be poor because of the length of the distribution chain and unsophisticated postharvest handling. This is an exciting product for restaurants and upscale markets that are looking for new and interesting produce that is not too unusual. Chinese broccoli is very cheap in Asian markets, but is costly at markets outside the Asian community.

Use. Chinese broccoli can be used in stir-fried and other cooked dishes. It is a versatile food, and is quick and easy to prepare. Prepare the vegetable as you would broccoli. It has a slightly peppery flavor, and a pungent, lightly bitter aftertaste.

Culture

Climatic requirements. A cool-season vegetable, Chinese broccoli should be grown the way you would grow broccoli.

Propagation and care. The plant seems to do well on plastic mulch if given adequate moisture. Plant seeds ½ inch deep and 1 inch apart. Thin to 6 inches apart in rows 12 to 24 inches wide. Do not prefertilize, since it is not a heavy feeder. The crop matures quickly and needs plenty of water.

Harvest and postharvest practices. Harvest timing is critical. Cut 8-inch stalks, including a few leaves,

Chinese broccoli packed for market. (Photo: Charlotte Glenn)

just before the flowers open. Each plant can be harvested several times, so check them frequently. Pressure-pack the harvested produce in 30-pound wire-bound lugs or standard waxed leaf cartons (1⅑ bushels). Thoroughly cool and hydrate the packed cartons or lugs (dunk them in cold water and put them on ice or into a cooler if you have it; otherwise get them quickly to the market).

The USDA storage recommendation is 32°F at 95 to 100% relative humidity, with a storage life of 10 to 14 days.

Sources

Seed

Nichols Garden and Nursery, 1190 North Pacific Highway, Albany, OR 97321

Sunrise Enterprises, P.O. Box 10058, Elmwood, CT 06110–0058

Tsang and Ma, P.O. Box 5644, Redwood City, CA 94063

More information

Harrington, Geri. 1984. *Grow your own Chinese vegetables.* Garden Way Publishing, Pownal, VT.

Mansour, N. S. 1990. *Chinese cabbage and leafy greens.* Vegetable Crops Recommendations, Oregon State University, Corvallis, OR.

Rubatzky, Vincent, and Mas Yamaguchi. 1997. *World vegetables, 2d ed.* Chapman and Hall, New York, NY.

Stephens, James. *Minor vegetables.* 1988. Cooperative Extension Bulletin SP-40, University of Florida, Gainesville, FL.

USDA. 1987. *Tropical products transport handbook.* Agric. Handb. 668. USDA, Washington, DC.

Prepared by Claudia Myers and David Visher.

Chinese Long Bean, Yard-Long Bean, Asparagus Bean

Vigna unguiculata subspecies *sesquipedalis* is a member of the Fabaceae (pea) family.

Chinese long bean is an annual climbing plant. A cousin to the cowpea or black-eyed pea, Chinese long bean is much more of a trailing, climbing variety, and often reaches heights of 9 to 12 feet. The pods grow to 12 to 30 inches long. The plant has large yellow to violet-blue flowers. Chinese long bean is an indeterminate plant: it continues to grow even after flowering and fruiting.

Other names. *Taao-hla-chao* (Hmong); *juro-kusasagemae* (Japanese); *dow gauk* (Chinese); *sitaw* (Filipino); *dau-dua* (Vietnamese).

Market Information

It is easy to make long beans look bad at market — old, dry beans look terrible and will sell poorly.

Current production and yield. Chinese long beans are available year-round from the Caribbean, Mexico, and California. Peak supplies are in the late summer or early fall. In a field test at Riverside, researchers obtained marketable yields of 7,500 to 11,100 pounds per acre with three different cultivars. Based upon the plants and their fruiting condition at the end of the harvest as well as the indeterminate nature of the crop, the potential yield was probably greater.

Three different-colored cultivars of long bean. (Photo: Charlotte Glenn)

Use. Pick the pods at maximum length but when they are still smooth, before the seeds mature or expand. At this tender stage, they can be snapped and cooked in various ways: stewed with tomato sauce; boiled and drained, and then seasoned with lemon juice and oil; or simmered in butter or oil and garlic. The pale green bean is meatier and sweeter than the dark green bean, which has a less delicate taste.

Culture

Climatic requirements. This warm-season crop can be planted in a wide range of climatic conditions, but is very sensitive to cold temperatures. It can tolerate heat, low rainfall, and arid soils, but the pods become short and fibrous with low soil moisture. Chinese long beans prefer high temperatures, conditions under which other green beans cannot be produced.

Long beans packed for the Los Angeles wholesale market. (Photo: Hunter Johnson)

Propagation and care. Plant seeds 1 to 2 inches deep in late spring when the soil is warm. Thin the plants to 6 to 12 inches in the row with 4 to 5 feet between rows. Since the long bean is a legume, some growers inoculate the seed with nitrogen-fixing *Rhizobium* bacteria rather than apply nitrogen fertilizer. For long beans, use the cowpea or "EL" strain of inoculant commercially available from various sources.

The plant's long, trailing growth habit requires a trellis for best production. Training the vines requires labor, about as much as for tomatoes or peas. The plant will climb by itself, but still needs some help and a very strong trellis system. The vines will grow to the top of your trellis, so don't build the trellis so high that it makes harvesting too difficult. Various trellising systems can be used. Chinese long beans will climb poles, especially if they are not entirely vertical and the poles are ¾ to 2½ inches in diameter, but the plants must be trained to the poles early in the season.

Fruits grow from open flower to marketable length in about 9 days. If the seed was not inoculated with a *Rhizobium,* high-nitrogen fertilizer may be required when seeding and during the growing season. A field test at Riverside supplied 3 inches of water and 10 pounds of nitrogen per acre per week. Long beans require more water than cowpeas.

Aphids, particularly the black bean aphid (*Aphis fabae*), are drawn to the pods of this plant. If you plant long beans near other crops that are infested with this aphid, you are asking for trouble. Thrips tend to be a pest early in the season, but the plants will often outgrow them, especially as the weather gets warmer and the plants grow faster. Mites can be a problem, primarily after insecticide applications, which often lead to mite outbreaks.

Harvest and postharvest practices. Harvested beans develop rusty patches quickly. Keep harvested beans moist while in coolers, since dehydration in the coolers will lower quality and may make them unmarketable. Old or wilted long beans are unmarketable. Sell the beans as fresh as possible.

The USDA storage recommendation is 40° to 45°F at 90 to 95% relative humidity, with an approximate storage life of 7 to 10 days.

Sources

Seed

Seeds Blüm, Idaho City Stage, Boise, ID 83706

Sunrise Enterprises, P.O. Box 10058, Elmwood, CT 06110–0058

Tsang and Ma, P.O. Box 5644, Redwood City, CA 94063

More information

Duke, James A. 1981. *Handbook of legumes of world economic importance.* Plenum Press, New York, NY.

Federal-State Market News Service. 1987. *San Francisco fresh fruit and vegetable wholesale market prices 1987 and 1988.* California Department of Food and Agriculture Bureau of Market News and USDA Marketing Service.

Harrington, Geri. 1984. *Grow your own Chinese vegetables.* Garden Way Publishing, Pownal, VT.

Knott, J. E., and J. A. Deanon. 1967. *Vegetable production in South-East Asia.* University of Philippines, College of Agriculture, Los Baños.

The Packer. 1989. *1989 Produce availability and merchandising guide.* Vance Publishing Corp., Overland Park, KS.

Rubatzky, Vincent, and Mas Yamaguchi. 1997. *World vegetables, 2d ed.* Chapman and Hall, New York, NY.

USDA. 1987. *Tropical products transport handbook.* Agric. Handb. 668. USDA, Washington, DC.

Prepared by Mark Van Horn and Claudia Myers.

Chive

***Allium schoenoprasum* is a member of the Amaryllidaceae (amaryllis) family.**

Chives are perennials easily distinguished by their growth in dense clumps, lack of well-formed bulbs, and ornamental-quality violet flowers. The tubular leaves are 6 to 10 inches long. No other onion has such a wide geographical distribution as the chive, and few species are more variable.

Other names: *Siu heung, tsung* (Chinese); *he* (Vietnamese).

Market Information

Current production and yield. Most chives are grown in Europe and North America. The California County Agricultural Commissioners have reported that chives were grown commercially in Alameda, Monterey, San Benito, and Santa Cruz counties. Chives are also grown in pots and gardens throughout the state.

Use. The slender, tubular, hollow green leaves are cut as needed for garnish and seasoning in salads, soups, and stews. Chives are also used as ornamentals, either in the garden or in household pots, because of the attractive rose to violet flowers that appear atop the uniform green clumps of leaves. Dried chives are increasing in popularity, and this is the intended market for most commercial fields. The flower stems are not palatable.

Culture

Climatic requirements. Chives are hardy in cold weather, withstand drought, and grow well in a wide variety of soils. They prefer moist, cool conditions.

Propagation and care. Readily raised from seeds, chives can also be propagated easily by dividing

The small hollow green leaves of chives are used for garnish and flavoring. The round purple flowers visible here and their stems are unpalatable. (Photo: Yvonne Savio)

the small rhizomes. The chive is a hardy plant, and requires little care. Plant the seed ½ inch deep every 2 to 4 inches in rows 8 inches apart. Thin the plants to one clump every 6 inches in row. Harvest with scissors or a sharp knife, cutting the leaves back to within 2 inches of the soil.

Planting, whether by seeding or by dividing, usually occurs in the spring or summer. The number of plants within a clump will double 5 or 10 times each year, given sufficient space, water, light, and nutrients. The plants grow most rapidly in spring and summer. Natural dormancy begins in the fall, and regrowth starts during the winter. Frequent watering and nitrogen applications are recommended for the spring and summer.

Sources

Seed

W. Atlee Burpee & Co., 300 Park Avenue, Warminster, PA 18974

The Cook's Garden, P.O. Box 65, Londonderry, VT 05148

Johnny's Selected Seeds, 299 Foss Hill Road, Albion, ME 04910

Le Jardin du Gourmet, P.O. Box 75, St. Johnsbury Center, VT 05863

Nichols Garden Nursery, 1190 North Pacific Highway, Albany, OR 97321

Park Seed Co., Cokesbury Road, Greenwood, SC 29647–0001

Seeds Blüm, Idaho City Stage, Boise, ID 83706

Shepherd's Garden Seeds, Shipping Office, 30 Irene Street, Torrington, CT 06790

Taylor's Herb Gardens, 1525 Lone Oak Road, Vista, CA 92084

Tsang and Ma, P.O. Box 5644, Redwood City, CA 94063

More information

Brewster, James, and Haim Rabinowitch. 1989. *Onions and allied crops.* CRC Press, Inc. Boca Raton, FL.

California Agricultural Statistics Service. 1987, 1988. *County Agricultural Commissioner data: 1987 and 1988 annual reports.* California Department of Food and Agriculture, Sacramento, CA.

Jones, Henry, and Louis Mann. 1963. *Onions and their allies.* Interscience Publishers Inc., New York, NY.

Stephens, James. *Minor vegetables.* 1988. Cooperative Extension Bulletin SP-40, University of Florida, Gainesville, FL.

Prepared by Claudia Myers and Ron Voss.

Cilantro, Chinese Parsley, Coriander

***Coriandrum sativum* is a member of the Apiaceae (parsley) family.**

Cilantro is an annual plant that grows 2 to 3 feet tall. Its leaves are light green, feathery, and flat. The distinctive flavor of cilantro leaves is quite different from that of parsley. While the leaves are used as an herb, the dried fruits, called coriander seed, are used as a spice and have an entirely different taste. Coriander seeds are hard, brownish yellow, spherical (⅛ inch diameter), and ribbed, and grow in symmetrical clusters. Cilantro is native to the eastern Mediterranean region and southern Europe.

Other names: *Joh tsu* (Hmong); *koendoro* (Japanese); *yuan sui* (Mandarin Chinese); *yim sai* (Cantonese Chinese); *yun tsai* (Chinese); *kinchi* (Filipino); *rao mui* (Vietnamese).

Market Information

Organically grown cilantro is a marketable product, since it is thought to have a better flavor than cilantro grown conventionally on a large scale.

Use. The leaves, known as cilantro or as Chinese or Mexican parsley, are used for flavor in Mexican salsas and Chinese dishes. When dried, the leaves lose their fragrance and flavor. The dried fruit, coriander seeds, are used whole or ground as a spice. The seed's aromatic essential oil is extracted and used to scent perfumes and cosmetics. Thai cuisine uses the root fresh and minced in salads and relishes.

Culture

Climatic requirements. Cilantro will grow in a wide range of conditions. The plants are sensitive to heat, and will bolt to seed quickly in warm weather. For continuous cropping, re-seed every 3 weeks through cool weather.

Propagation and care. Coriander is a hard seed and may need to be cracked or scarified before planting. Plant the seed ½ inch deep every 2 inches in rows 12 inches apart. Thin to one plant every 6 to 8 inches. If you harvest the older, outside leaves, the plant will continue to produce new foliage until it goes to seed. Large-scale commercial growers clip the plant just below ground level and bunch it. Some growers cut it off 1 inch above the ground, thus allowing the plant to regrow for a second cutting.

Fresh cilantro leaves displayed for market. (Photo: Alf Christianson Seed Co.)

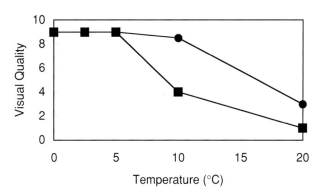

Chart: Effect of holding temperature on the quality of fresh cilantro after 1 (●) and 2 (■) weeks' storage in perforated polyethylene bags. Visual quality was assessed on a 5-point scale (1, 3, 5, 7, 9, where 1 = low quality and 9 = high quality). From Joyce, Reid, and Katz 1986.

Harvest the coriander seed when the entire plant is dried and crisp, but before the seed pods break open and scatter seed. Cut the whole plant, threshing it out for further drying, or hang it to dry, gathering the seed as it falls. Undried seed has a bitter taste.

Disease Problems

Bacterial leaf spot. At least as early as 1988 a leaf spot disease appeared on cilantro in Southern California. 1990 and 1991 saw moderate to severe infections of this disease in Monterey, Santa Cruz, and other central coast counties. The disease has been identified as bacterial leaf spot caused by the pathogen *Pseudomonas syringae*.

Symptoms consist of angular, vein-delimited leaf lesions that are at first water-soaked or translucent. The spots characteristically penetrate the entire leaf; that is, any one particular spot will be visible on the top and bottom of the leaf. Over time and with drying conditions, the leaf spots may turn black or brown. If infection is severe, leaf spots may coalesce and cause a blighting effect. Researchers in England state that some stunting and yellowing are also associated with this disease, but we are not certain these symptoms have occurred on infected cilantro in California.

Very little information is available on the development of *P. syringae* on cilantro. The disease has been reported only in England, Hungary, and California. Under experimental conditions the pathogen will also infect parsley. Of particular note is the pathogen's seedborne transmission. Infected or infested cilantro seed is an important means by which the disease spreads and establishes itself.

Very few control options are currently available to growers. In attempting to manage this disease, growers should consider the following:

1. Splashing water enhances disease development and spread, so rain and sprinkler irrigation favor the pathogen.

2. Infested or infected seed is probably the main way the pathogen enters fields. Pathogen-free seed is therefore very important, but researchers have not yet developed a test to detect the pathogen in seed.

3. Seed treatments using antibiotics gave effective control of the disease in experiments in England, but no such treatments are registered in California. No data are available on the effect of hot-water treatments of cilantro seed.

4. We have no data on the efficacy of bactericides, nor are any such materials registered for use on cilantro in California.

Carrot motley dwarf. Several plants in the Apiaceae family, including carrot and cilantro, are affected by the viral disease carrot motley dwarf. Symptoms on cilantro consist of bright yellow and red coloration of the normally green foliage. Some stunting of the plants may be observed. The virus is vectored only by the carrot willow aphid and has been confirmed strictly in the coastal regions of the state. Plant pathologists are currently researching the nature of the virus pathogen and disease epidemiology. No control measures are suggested at this time.

Postharvest Handling

Use only the highest-quality plant material for the fresh market. Some of the detailed postharvest handling information provided for basil also applies to cilantro and other herbs.

Sources

Seed

NOTE: Cilantro seed is widely available, and the seed sold in supermarkets that carry Mexican spices will probably grow just fine. However, some seed intended for use as a seasoning may be unusable for propagation, whether because of heating, other processing, or age.

More information

Federal-State Market News Service. 1988, 1989. *Los Angeles fresh fruit and vegetable wholesale market prices 1988 and 1989.* California Department of Food and Agriculture Bureau of Market News and USDA Marketing Service.

Kowalchik, Claire, et al., eds. 1987. *Rodale's illustrated encyclopedia of herbs.* Rodale Press, Emmaus, PA.

Organic Gardening Magazine. 1978. *The encyclopedia of organic gardening.* Rodale Press, Emmaus, PA.

Prepared by Keith Mayberry, Yvonne Savio, Claudia Myers, and Steven Koike (disease section).

Citron, Preserving Melon

Citrullus lanatus (***C. vulgaris***) var. ***citroides*** is a member of the Cucurbitaceae (gourd) family. Don't confuse this with the other, better-known fruit also called citron; that fruit, ***Citrus medica***, grows on a small tree and looks like a lemon.

Citron or preserving melon is a close relative of watermelon, but is inedible raw. The flesh and rind are sometimes used for making preserves and pickles. The fruit resembles a watermelon, and is round to oval, 6 to 12 inches long, and light green with darker green stripes and a smooth surface. The flesh is typically white, although some forms have light green or pink flesh. Citron fruit is so tough that you can bounce it on the floor with only a small chance that it will burst. The plant is a low, spreading vine. The plant has both male and female flowers; bees are required for pollination. It is native to Africa.

Citron grows as a weed in California and can infest annual crops, orchards, vineyards, roadsides, ditchbanks, and sandy, dry river or creek beds. It is a troublesome weed in Imperial Valley asparagus fields. Where watermelons are grown, citron becomes particularly undesirable since the two plants readily hybridize.

Citron is a close relative of watermelon, but is inedible raw. Mature fruit are 6 to 12 inches long. (Photo: Hunter Johnson)

Market Information

Use. Citron flesh and rind are sometimes used for sweet preserves and glacé fruit. The fruit is occasionally used as hogfeed. Commercial production is rare.

Sources

Seed

Le Jardin du Gourmet, P.O. Box 75, St. Johnsbury Center, VT 05863

Seeds Blüm, Idaho City Stage, Boise, ID 83706

More information

Rubatzky, Vincent, and Mas Yamaguchi. 1997. *World vegetables, 2d ed.* Chapman and Hall, New York, NY.

Stephens, James. *Minor vegetables.* 1988. Cooperative Extension Bulletin SP-40, University of Florida, Gainesville, FL.

University of California. 1996. WI-131 citron, in *Grower's weed identification handbook.* Publication 4030, UC Division of Agriculture and Natural Resources, Oakland, CA.

Prepared by Claudia Myers (adapted from James Stephens's Minor Vegetables*).*

Collards

Brassica oleracea, Acephela group is a member of the Brassicaceae (mustard) family.
Varieties include Vates, Champion, Georgia, Morris Heading, and Louisiana Sweet.

The collard is a green leafy vegetable. The dark green leaves are borne in rosettes around an upright, stocky main stem. The long-stemmed leaves resemble cabbage leaves, except that they are oval rather than round. It is a common cooked green in the southern United States.

Market Information

Leaves can be clipped from the plant as long as the weather and the plant's flavor hold up. A good freeze will greatly improve flavor.

Bunch the harvested leaves tightly, tie them, pack them, and hydrocool them. Pack 24 bunches in a 25-pound wirebound or leaf carton. Sell by offering samples of fresh leaves.

Current production and yield. In California, the County Agricultural Commissioners have reported that collard greens were grown commercially in Fresno, San Diego, and Santa Barbara counties.

In the United States, collards are available in the market year round, but supplies peak between December and April and decline from June through August. Virginia, Georgia, Florida, South Carolina, and Alabama are the leading producers of collards.

Use. The plant's nutritious leaves can be boiled or stir-fried as greens. They have a slightly bitter flavor.

Nutrition. Collard greens are high in Vitamin A, with about twice the Vitamin A of broccoli. A 100 g edible serving contains 3,300 IU of Vitamin A, 23 mg of Vitamin C, 1.6 g of protein, and 117 mg of calcium.

Culture

Climatic requirements. Collards thrive over a wide range of growing conditions, but the quality and taste are better and the plant grows best during the cooler months of the year. Along the coast they can be grown all summer. The plants can withstand temperatures as low as 15°F unless such a freeze abruptly follows a warm period of growth. They withstand heat well, and can take more cold than cabbage.

Collard leaves grow in a rosette around an upright stem. (Photo: Hunter Johnson)

Propagation and care. In commercial plantings, collards are direct-seeded ½ inch deep. Space the rows 24 to 36 inches apart, four seeds per foot, and then thin to one plant every 10 to 12 inches. About 2 to 4 pounds of seed will plant an acre. If you transplant, space the plants 10 inches apart in the rows.

Collards grow very well in well-drained loam soils that are high in organic matter. About 6 to 8 weeks after seeds are planted, collards are ready to harvest. Harvest can continue through several pickings as long as the weather is cool and leaf quality remains good. Thrips, aphids, flea beetles, and cabbage loopers are common pests of collards.

Harvest and postharvest practices. Handle collard greens as you would handle spinach. The USDA storage recommendation is 32°F at 95 to 100% relative humidity to prevent wilting, with an approximate storage life of 10 to 14 days.

Sources

Seed

Asgrow Seed Co., P.O. Box 5038, Salinas, CA 93915

W. Atlee Burpee & Co., 300 Park Avenue, Warminster, PA 18974

Ferry-Morse Seed Co., P.O. Box 4938, Modesto, CA 95352

Harris Moran Seed Co., 3670 Buffalo Road, Rochester, NY 14624

Johnny's Selected Seeds, Foss Hill Road, Albion, ME 04910

Le Jardin du Gourmet, P.O. Box 75, St. Johnsbury Center, VT 05863

Nichols Garden and Nursery, 1190 North Pacific Highway, Albany, OR 97321

Park Seed Co., Cokesbury Road, Greenwood, SC 29647–0001

Seeds Blüm, Idaho City Stage, Boise, ID 83706

NOTE: Also, check with your local seed suppliers.

More information

California Agricultural Statistics Service. 1987, 1988. *County Agricultural Commissioner data. 1987 and 1988 annual reports.* California Department of Food and Agriculture, Sacramento, CA.

Federal-State Market News Service. 1989. *Los Angeles fresh fruit and vegetable wholesale market prices 1989.* California Department of Food and Agriculture Bureau of Market News and USDA Marketing Service.

Hall, Harwood, Susan Wada, and Ron Voss. 1976. Growing cole crops. Leaflet 2927, UC Division of Agriculture and Natural Resources, Oakland, CA.

Mansour, N. S. 1990. *Collards.* Vegetable Crops Recommendations. Oregon State University, Corvallis, OR.

The Packer. 1990. *1990 produce availability and merchandising guide.* Vance Publishing Corp., Overland Park, KS.

Stephens, James. *Minor vegetables.* 1988. Cooperative Extension Bulletin SP-40, University of Florida, Gainesville, FL.

USDA. 1987. *Tropical products transport handbook.* Agric. Handb. 668. USDA, Washington, DC.

USDA. 1984. *Vegetables and vegetable products.* Agric. Handb. 8. USDA, Washington, DC.

Prepared by Claudia Myers (adapted from James Stephens's Minor Vegetables*).*

Daikon, Lobok, Oriental or Chinese Radish

Daikon, *Raphanus sativus,* **is in the Longipinatus group, a member of the Brassicaceae (mustard) family. Varieties include Mino Spring Cross (spring cultivar), Summer Cross No. 3, Chinese White (cylindrical), Chinese Rose (round), Celestial (cylindrical), and Tokinashi (a good all-season cultivar).**

Oriental radishes have extremely large roots. The Sakurajima variety, one of the largest, can weigh as much as 50 pounds. Most oriental radishes are in the 1 to 2½ pound class at full maturity. Leaf spreads of more than 2 feet are common. The leaves differ from those of spring radish types in that they have great notches and they spread from the root tops in rosette fashion. Some varieties form large round or top-shaped roots, while others are cylindrical.

Other names. *Lo pue* (Hmong); *daikon* (Japanese); *lor bark* (Cantonese Chinese); *labanos* (Filipino); *cu-cai trang* (Vietnamese).

Market Information

Use. This type of radish is usually cooked rather than eaten fresh, but it can be used raw in salads. In Japan, the radishes are often pickled.

Nutrition. Daikon contains no Vitamin A, 22 mg of Vitamin C, 0.6 g of protein, and 27 mg of calcium per 100 g raw, edible portion (about 1 cup of sliced daikon).

Culture

Propagation and care. Culture is similar to that used for more familiar radish varieties. Plant the seed ¾ inch deep in the fall (September through October) so the roots will enlarge in the cool months. The seeds and then the plants should be 4 to 6 inches apart in rows 3 feet apart. To compensate for the large root size, plant oriental radishes on high raised beds fortified with organic matter (compost). At each cultivation, work the soil higher and higher around the root as it grows. Most oriental radishes reach their best useable size in 60 to 70 days.

Daikon packed for the Los Angeles wholesale market. The roots are 6 to 20 inches long. (Photo: Hunter Johnson)

The oriental radish on display at the Los Angeles wholesale market. (Photo: Hunter Johnson)

Harvest and postharvest practices. The USDA storage recommendation are 32° to 34°F at 95 to 100% relative humidity, with an approximate storage life of 4 months.

Sources

Seed

W. Atlee Burpee & Company, 300 Park Avenue, Warminster, PA 18974

Johnny's Selected Seeds, Foss Hill Road, Albion, ME 04910

Nichols Garden Nursery, 1190 North Pacific Highway, Albany, OR 97321

Park Seed Co., Cokesbury Road, Greenwood, SC 29647–0001

Seeds Blüm, Idaho City Stage, Boise, ID 83706

Shepherd's Garden Seeds, Shipping Office, 30 Irene Street, Torrington, CT 06790

Sunrise Enterprises, P.O. Box 10058, Elmwood, CT 06110–0058

Tsang and Ma, P.O. Box 5644, Redwood City, CA 94063

More information

Forsythe, Adrian. 1987. Of radishes and kings. *Harrowsmith,* May-June 1987, pp. 90–93

Federal-State Market News Service. 1988, 1989. *Los Angeles fresh fruit and vegetable wholesale market prices 1988–1989.* California Department of Food and Agriculture Bureau of Market News and USDA Marketing Service.

Federal-State Market News Service. 1987, 1988. *San Francisco fresh fruit and vegetable wholesale market prices 1987–1988.* California Department of Food and Agriculture Bureau of Market News and USDA Marketing Service.

Harrington, G. 1978. *Grow your own Chinese vegetables.* Garden Way Publishing, Pownal, VT.

Rubatzky, Vincent, and Mas Yamaguchi. 1997. *World vegetables, 2d ed.* Chapman and Hall, New York, NY.

Stephens, James. *Minor vegetables.* 1988. Cooperative Extension Bulletin SP-40, University of Florida, Gainesville, FL.

USDA. 1987. *Tropical products transport handbook.* Agric. Handb. 668. USDA, Washington, DC.

Prepared by Claudia Myers.

Dill

**Anethum graveolens is a member of the Apiaceae (parsley) family.
Varieties include Bouquet and Mammoth.**

Dill, an annual of the parsley family, is indigenous to the Mediterranean region and southern Russia. The medium-sized herb reaches 3 feet in height, and has small, feathery, blue-green leaves, yellow flowers, and hollow stems. One long, hollow stalk comes up from the root. The fruit (or seeds) are ribbed, flattened, elliptical, and about ⅙ inch long. Dill is related to anise, caraway, coriander, cumin, fennel, and parsley.

Market Information

Current production and yield. India and Pakistan are the principal world producers of dill. Egypt, Fiji, Mexico, the Netherlands, the United States, England, Hungary, Germany, and Holland also have commercial production areas.

Use. Both the dried seeds and the fresh, feathery leaves or fronds (called dillweed) are used in cooking. Dill is well known for its use in pickles and in Scandinavian and German food, although the plant is indigenous to southern Europe. The ancient Greeks and Romans made great use of dill, and fresh dill is still used widely in Greece, especially in the winter. The herb is very pungent. The dried seeds lose some of their flavor, however, and are slightly reminiscent of caraway — hardly surprising, since they share the same essential oil. Use dill seeds and dillweed with pickles, salad dressings, potatoes, tomatoes, cream, cucumbers, cream cheese, fish, lamb, and string beans.

Dill reaches a height of 3 feet and has small, feathery, blue-green leaves, yellow flowers, and hollow stems. (Photo: Charlotte Glenn)

Packaging. Dill is packaged 12 to 18 plants to the bunch for fresh market sales.

Culture

Climatic requirements. Dill is a hardy plant that thrives on long days and cool weather. It is sensitive to stresses such as insufficient water, hail, high temperatures, strong winds, and hard rains during flowering.

Propagation and care. The reported life zone is 45° to 80°F with an annual rainfall of 20 to 60 inches and a soil pH of 5.3 to 7.8. Propagate by seed in spring or autumn. If planted in the spring, germination will takes 10 days to 2 weeks. If planted in the late fall, germination won't take place until the following spring. Plant in light, well-drained soil (preferably deep, fertile loam soils) in a sunny, sheltered location. Plant 15 to 20 seeds per foot of row ¼ inch deep, and thin to 3 or 4 plants per foot with rows 1 to 3 feet apart. One or two pounds of seed are needed per acre for dill grown for fresh market.

Diseases generally do not present problems. Insect pests include aphids, including "carrot aphids," green plant lice that colonize the dill heads. Seedcorn maggot is a small, legless maggot that feeds on and destroys germinating seeds. Proper rotations and field selection can minimize insect problems.

Dill competes poorly with weeds. Cultivate as often as necessary when weeds are small. Proper cultivation, field selection, and rotations can reduce or eliminate the need for chemical weed control. Irrigate during any periods of growth. Wind can destroy the tall, spindly stalks. The seeds ripen in autumn and can be collected as soon as the first few fall, 2 to 3 weeks after blossoming. Snip off the flower heads and spread them out on trays in the sun for a few days. When the flower heads are completely dry, the seeds will release easily from the heads. Store in airtight containers.

Fresh dillweed can be refrigerated for only a couple of days before it droops and loses flavor. When drying the aromatic, anise-tasting leaves, start cutting the leafy stalks before the flower heads appear. Spread the leaves on a wire rack in a shady, cool place.

In large-scale production, shattering losses can be a problem. Seeds tend to ripen at different times, so the timeliness of harvest is critical to maximize seed yield. If you are growing dill for its essential oil, harvest it before it flowers and then steam-distill it the same day.

Sources

Seed

NOTE: Dill seed is widely available.

More information

Kowalchik, Claire, and William Hylton, eds. 1987. *Rodale's illustrated encyclopedia of herbs.* Rodale Press, Emmaus, PA.

Mansour, N. S. 1990. *Dill.* Vegetable Crops Recommendations. Oregon State University, Corvallis, OR.

Simon, James, Alena Chadwick, and Lyle Craker. 1984. *Herbs: An indexed bibliography 1971–1980.* Archon Books, Hamden, CT.

Prepared by Curt Robinson and Claudia Myers.

Endive, Escarole, Chicory

Cichorium endivia **is a member of the Asteraceae (sunflower) family. Varieties include Green Curled Ruffec (endive), Broadleaf Batavian (escarole), and Full Heart Batavian (escarole).**

Endive is a loose-headed plant with narrow, curling leaves. The broadleaf type is called escarole. Full Heart Batavian is the main variety of escarole on the market. At harvest, the rosette of leaves makes up most of the plant. It is 12 to 15 inches across with upright to spreading growth and deep green leaves.

Market Information

Current production and yield. Florida is the leading U.S. producer of endive and escarole. Other producing states include California, New Jersey, Ohio, and New York. Supplies are available year-round, and peak from December through April.

Use. The leaves of endive and escarole are used as salad and cooked greens. A 2-cup portion of either of these vegetables contains about 2,050 IU of Vitamin A, 6.5 milligrams of Vitamin C, 1.25 grams of protein, and 52 milligrams of calcium.

Culture

Climatic requirements. Endive and escarole are hardy plants but do poorly in hot weather and are damaged by severe frosts. Seed them to mature before hot weather arrives. Escarole is the more cold-tolerant of the two. Both plants thrive best as early spring or late fall crops. The plants bolt in hot weather.

Propagation and care. Endive is similar to lettuce in its soil and climatic requirements. Plant 1 foot apart in rows spaced 1½ or 2 feet apart. Blanching the head for two weeks will reduce bitterness. To blanch, tie the head like cauliflower or stand boards on each side of the row. Tie the leaves together only when they are dry, since wet foliage during blanching will decay. Blanching is not a common practice among commercial growers.

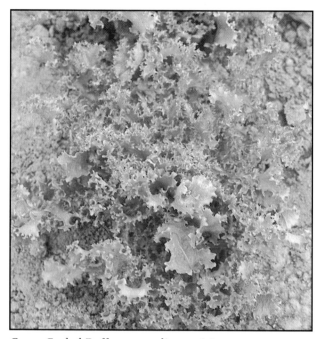
Green Curled Ruffec, an endive variety. (Photo: Hunter Johnson)

Harvest and postharvest practices. The USDA recommends storing at 32°F with 95 to 100% relative humidity. The approximate storage life is 2 to 3 weeks.

Sources

Seed

W. Atlee Burpee & Co., 300 Park Avenue, Warminster, PA 18974

The Cook's Garden, P.O. Box 65, Londonderry, VT 05148

Johnny's Selected Seeds, Foss Hill Road, Albion, ME 04910

Le Jardin du Gourmet, P.O. Box 75, St. Johnsbury Center, VT 05863

Nichols Garden Nursery, 1190 North Pacific Highway, Albany, OR 97321

Park Seed Co., Cokesbury Road, Greenwood, SC 29647–0001

Seeds Blüm, Idaho City Stage, Boise, ID 83706

Shepherd's Garden Seeds, Shipping Office, 30 Irene Street, Torrington, CT 06790

Taylor's Herb Gardens, 1525 Lone Oak Road, Vista, CA 92084

More information

California Agricultural Statistics Service. 1980–1988. *County Agricultural Commissioner data. 1980–1988 annual reports.* California Department of Food and Agriculture, Sacramento, CA.

Chandoha, Walter. 1984. Grow Italian greens—radicchio, escarole, and arugula. *Organic Gardening,* 31(5):80–84.

Economic Research Service. 1989. *Vegetables and specialties: Situation and outlook yearbook.* USDA, Washington, DC.

Federal-State Market News Service. 1988, 1989. *Los Angeles fresh fruit and vegetable wholesale market prices 1988 and 1989.* California Department of Food and Agriculture Bureau of Market News and USDA Marketing Service.

Federal-State Market News Service. 1987, 1988. *San Francisco fresh fruit and vegetable wholesale market prices 1987 and 1988.* California Department of Food and Agriculture Bureau of Market News and USDA Marketing Service.

Kline, Roger. 1986. Specialty crops will fill a market niche. *American Vegetable Grower,* April 1986, pp. 29–37.

Mansour, N. S. 1990. *Endive and escarole.* Vegetable Crops Recommendations. Oregon State University, Corvallis, OR.

The Packer. 1989. *1989 produce availability and merchandising guide.* Vance Publishing Corp., Overland Park, KS.

United Fresh Fruit and Vegetable Association. 1964. Endive-escarole-chicory. *Fruit & Vegetable Facts & Pointers.* September 1964.

USDA. 1984. *Composition of foods: Vegetables and vegetable products.* Agric. Handb. 8–11. USDA, Washington, DC

USDA. n.d. *Table of container net weights.* USDA Marketing Service, Washington, DC.

Prepared by Claudia Myers.

European Black Currant

Ribes nigrum is a member of the Saxifragaceae (saxifrage) family.

The European black currant (*Ribes nigrum*) is a deciduous shrub native to northern Europe and to north and central Asia. It has stiff, upright branches that grow 5 to 6 feet tall. Flowers are borne on 1-year-old wood and on tiny spurs on older wood. Most cultivars do not require cross-pollination, but the flower structure of a few varieties prevents ready self-pollination. Black currant fruits are born in strigs, or chains, similar to but shorter than those of the red or white currant. Fruit size averages about ⅜ inch in diameter.

Because so many share a common ancestry, European black currant cultivars do not vary greatly in plant or fruit characteristics. Growers are now beginning to cross *Ribes nigrum* with other *Ribes* species in an attempt to introduce certain desirable characteristics. *Ribes bracteosum* crosses such as Jet have longer strigs of berries, *R. nigrum sibirica* and *R. dikuscha* crosses are more cold-hardy and resistant to disease, and *R. ussuriense* crosses like Consort are resistant to white pine blister rust.

Two other black currant species that have occasionally been cultivated for their fruits are the clove currant (*R. odoratum*) and the American black currant (*R. americanum*). Around 1900, clones were selected with superior fruit, but neither species is widely grown today.

European black currants (and other true currants, for that matter) are not related to the small raisins that have long been sold commercially as "dried currants" or "black currants," and which are in fact dried Black Corinth grapes.

Market Information

The black currant is popular in northern Europe for its flavor and its high Vitamin C content. The plant and fruit are little known in America. The plant was recognized as a host for white pine blister rust, and banned from the United States in the 1920s. The ban was lifted in 1966, and American interest in rust-resistant varieties of this fruit is now growing.

Use. European black currant has a strong flavor. People who enjoy the flavor eat the berries out of hand. The best cultivars for fresh eating include

These full-sized, firm European black currants are ready for harvest. (Photo: Bernadine Strik)

Blackdown, Brodtorp, Goliath, and Silvergieters. The fruit is also used in juice, jams, tarts, and wines.

Culture

Climatic requirements. Black currants tolerate cold temperatures to –25°F or lower, depending on the cultivar. They do not, however, tolerate extremely hot summers, particularly in conjunction with dry weather.

Propagation and care. Black currants thrive in cool, well-drained, fertile soils. The bush will thrive in full sun or in partial shade. In warm-summer areas, plant the bushes in partial shade or on a north-facing slope. In warm regions, bushes produce better in heavier soils that retain moisture and remain cooler. An organic mulch helps to protect the shallow roots.

Slightly acidic soils that are rich in organic matter are best. Apply nitrogen in a yearly mulch of strawy manure or fertilizer that can supply about 4 ounces of nitrogen per square yard. Potassium is needed at the rate of ½ ounce of actual potassium to the square yard.

Currants leaf out early in spring, so plant either very early in the spring or in the fall, with a mulch to prevent heaving during the winter. Set the transplants slightly deeper than they grew in the nursery. This will encourage buds and new shoots at and below ground level. You can cut off all branches to within 1 inch of the ground after planting to encourage plants to grow strong shoots and roots. The leaf canopy may develop more quickly and provide more vigor, however, if 1-year-old canes are left on the transplants. If you are planting European black currants as individual bushes, space the plants 6 feet apart; space them only 3 feet apart in a row if you want to grow them as a hedge.

Black currant bushes need annual pruning. When you prune, leave an adequate supply of 1-year-old wood to bear fruit and stimulate new shoot growth for fruit the following season. In the winter after the first growing season, prune all but six of the strongest upright shoots. In subsequent winters, cut away old shoots at ground level or shorten them to vigorous side shoots. Also remove branches that are broken, trailing on the ground, or diseased.

If currant production declines with plant age, cut off all the branches at ground level in winter. Though this will sacrifice the next season's crop, you should get a good load of European black currants the season after that on vigorous 1-year-old wood.

Propagation. Hardwood cuttings will root readily if each 8- to 12-inch piece is set in the ground so only the top bud is exposed. Set the cuttings in early spring or autumn. They can even be set at the end of the summer if you leave the top leaves attached and do not let plants dry out before rooting. Softwood cuttings also root easily. Three-inch tip cuttings, given shade and either mist or a clear plastic tent, will grow roots in 3 or 4 weeks.

Drooping branches of black currants often layer themselves (take root). If you only want one or two new plants, you can encourage layering by bending a low branch to the ground and covering it with some soil and a stone.

Harvest. Pick black currants while they are dry and still firm. Take the whole strig unless the fruit is to be used immediately. If you are picking for fresh eating, make sure the berries are fully ripe.

Pests and diseases. European black currant is a host for white pine blister rust, but usually is not seriously affected by the disease. Some resistant (non-host) cultivars have been developed. Some European black currant cultivars are susceptible to mildew and leaf spotting diseases, but often these diseases can be controlled by choosing the right cultivar and an auspicious planting site, and by practicing proper pruning.

Sources

Plants

Alexander Eppler Ltd., P.O. Box 16513, Seattle, WA 98116-0513

Edible Landscaping Nursery, Rte. 2, Box 77, Afton, VA 22920

International Ribes Association, c/o Anderson Valley Agriculture Insitute, P.O. Box 130, Boonville, CA 95415

Raintree Nursery, 391 Butts Rd., Morton, WA 98356

Tolowa Nursery, 360 Stephen Way, Williams, OR 97544

Whitman Farms Nursery, 1420 Beaumont NW, Salem, OR 97304

More information

Antonelli, A., et al. 1988. *Small fruit pests: Biology, diagnosis, and management.* Publication EB 1388, Washington State University Agricultural Communications, Pullman, WA.

Baker, Harry. 1986. *The fruit garden displayed.* Cassell Ltd., The Royal Horticultural Society, London.

Card, Fred Wallace. 1898. *Bush fruits.* Macmillan, London.

Darrow, G., and S. Detwiler. 1924. *Currants and gooseberries: Their culture and relation to white-pine blister rust.* Farmer's Bulletin No. 1398. USDA, Washington, DC.

Galletta, G., and D. Himelrick, eds. 1990. *Small fruit crop management.* Prentice Hall Press, West Nyack, NY.

Ourecky, D. K. 1977. *Blackberries, currants, and gooseberries.* Cooperative Extension Publication IB 97. Cornell University, Ithaca, NY.

Reich, Lee. 1991. *Uncommon fruits worthy of attention: A gardener's guide.* Addison-Wesley Publishing Co., Reading, MA.

Prepared by Lee Reich.

Fava Bean

Vicia faba L. is a member of the Fabaceae (pea) family.

Fava beans are related to vetch, a determinate plant with erect, coarse stems, large leaves, and no climbing tendrils. It is a bushy plant, 2 to 7 feet tall, with white or purplish flowers born in clusters on short stalks. Large-seeded varieties bear 1 or 2 pods at each node and small-seeded types produce 2 to 5 pods. The pods are up to 18 inches long and contain from 3 to 12 large beans. There are about 15 pods per stalk on the large types, and 60 pods on plants of the small-seeded varieties. The bean is similar in size to the Lima bean.

Other names. The fava bean is also called horse bean, broad bean, Windsor bean, English dwarf bean, tick bean, pigeon bean, bell bean, haba bean, feve bean, and silkworm bean.

Market Information

For the past several years, fresh fava beans have been listed in the San Francisco Wholesale Fruit and Vegetable Report only during the months of April and May.

Current production. In North America, Canada is the largest producer of fava beans. Minnesota and the Great Lakes states grow small acreages. In California, fava beans are grown as a seed crop along the coast from Lompoc to Salinas and in the northern Sacramento Valley. In other parts of the state they are grown primarily as a cover crop or for green manure.

Use. Fava beans are used as a green or dry vegetable and as a coffee extender when roasted and ground. As a food, the beans are hulled, boiled, and prepared in the same way as lima beans. Young, fresh fava beans can be cooked without hulling. The plant is also used as livestock and poultry feed and as a cover crop.

Nutrition. The dry beans are about 24 percent protein, 2 percent fat, and 50 percent carbohydrate, and have 700 calories per cup.

Culture

Climatic requirements. The fava bean is a cool-season annual legume. Optimal growing temperatures range from 70° to 80°F. The plant resists frost damage at least to 21°F, but does poorly in the summer heat of interior valleys, especially during flower and pod set. In San Joaquin County, November plantings have tolerated winter frost well, and seed from those plantings have been harvestable in April.

Propagation and care. In California, fava beans are planted for vegetable use in February and March. As cover crops, they are planted from September to November. When grown for seed production, the crop takes 4 to 5 months to mature. The seeds should be planted 1 to 2 inches deep (large varieties) into well-prepared soil, 3 to 5 inches apart. Germination takes 7 to 14 days. The sprouted seeds are thinned to 8 to 10 inches apart if practical. Allow 2 to 3 feet between rows for seed production.

Mature fava beans, ready for harvest. (Photo: Sakata Seed America, Inc.)

For cover crops, one plant per square foot is recommended. This is equivalent to 195 pounds of seed per acre for large-seeded varieties and 79 pounds of seed per acre for the small-seeded varieties.

Plants do best on well-drained heavy silt or clay loams, but will also do well on adequately moistened sandy soils. Fava beans are relatively tolerant to boron (up to 10 ppm in irrigation water). Fava beans do not need nitrogen fertilizer if the plants are properly nodulated. Inoculation with commercially available legume-type bacteria is recommended. Yields of 1 to 2 tons of cleaned seed per acre can be expected.

Fava bean is attractive as a green manure crop because it produces a large amount of biomass. Twenty to forty tons of green forage worked into each acre of soil enhances the tilth of many clay and sandy soils. Fava bean adds a large amount of nitrogen to the soil, and that benefits orchards or high-nitrogen-consuming annual crops. Green manure yields were determined in several research trials throughout the state have ranged from 4.9 tons per acre to 34 tons per acre.

Seed production. California research has shown small-seeded varieties such as Bell and Foul to yield around 2,400 pounds per acre. Large-seeded varieties—Burpee, Pismo, and Italy—yield around 1,900 pounds of seed per acre.

The nutrient composition of fava bean green material is about ½ pound of nitrogen for every 100 pounds of material turned under. A green material yield of 20 tons per acre would be equivalent to approximately 200 pounds of nitrogen.

Pests and diseases. Fava beans are susceptible to aphid and bean weevil. Ladybird beetles and some small parasitic wasps can be effective in controlling aphids. Little is known about diseases affecting fava beans under California conditions. Chocolate spot (*Botrytis fabae*), a fungus disease that causes brown spotting under moist conditions, has been identified in the northern San Joaquin Valley. Viruslike diseases and powdery mildew have been observed on fava beans. Fava beans are a preferred host of black aphids, and may serve as a bridging host for virus diseases that the aphid transmits to other legumes or susceptible crops.

Harvest and postharvest practices. Select the pods when they are green and thick and have a glossy sheen. These pods should be well filled with large beans. The raw beans can then be kept in the refrigerator for a day or two.

Swathing should begin when the lowest two bunches of pods begin blackening or when most seeds easily detach from the hilium. At this stage the moisture content of the beans is from 35 to 45 percent.

Favism

Favism is an inherited disorder particularly noted among persons of southern European heritage. It is an enzyme deficiency expressed when fava beans, especially in raw or partially cooked form, are eaten. Symptoms include acute toxic hepatitis as well as symptoms similar to those of influenza. Males are more commonly affected than females; mortality from favism is almost entirely confined to children. Fava plant pollen in the respiratory tract also affects these people.

Sources

Seed

Harmony Farm Supply, P.O. Box 460, Graton, CA 95444

Lockhart Seeds, Inc., P.O. Box 136, Stockton, CA 95201

Mellinger's, 2310 W. South Range Rd., North Lima, OH 44452–9731

Seeds Blüm, Idaho City Stage, Boise, ID 83706

More information

Doty, Walter L. 1980. *All about vegetables.* Ortho Books, San Francisco, CA.

Kennedy, P. B. 1923. *Small-seeded horse bean.* Circular 257, UC Agricultural Extension Service, Berkeley, CA.

Kingsbury, J. M. 1964. *Poisonous plants of the U.S. and Canada.* Prentice Hall Press, West Nyack, NY.

Oplinger, E. S., et al. 1989. *Fava bean.* Alternative Field Crops Manual, University of Wisconsin Extension.

Rubatzky, Vincent, and Mas Yamaguchi. 1997. *World vegetables, 2d ed.* Chapman and Hall, New York, NY.

Smith, F. L. 1957. *Three new varieties of fava beans.* UC Cooperative Extension, San Mateo County, CA.

Sunset. 1995. *Sunset western garden book.* Sunset Publishing Corp., Menlo Park, CA.

Prepared by Gary Hickman and Mick Canevari.

Fennel, Sweet Anise

Foeniculum vulgare or *F. officinale* (usually called common or wild fennel) and *Foeniculum vulgare dulce* (called Florence or sweet fennel or finocchio) are members of the Apiaceae (parsley) family.

Fennel is a tall, hardy, aromatic perennial of the parsley family native to southern Europe and the Mediterranean area, especially near the sea. It is distinguished by its finely divided, feathery green foliage and its golden-yellow flowers. It looks something like celery and is often confused with dill. The spice, fennel "seed," is really the dried fruit of the common fennel.

There are several varieties of fennel. Both common and sweet fennels are grown for their seed and their essential oils. The seed is used as a spice in cooking and the essential oil is used in condiments, soaps, creams, perfumes, and liqueurs. Sweet fennel (Florence or finnochio fennel) is also grown for the thickened bulblike base of the leaf stems (often called anise), a 3- or 4-inch-wide structure that grows just above the ground. Fennel plants grown for this purpose grow only 2 to 3 feet tall. Seeding types will grow 4 to 5 feet tall.

Many varieties (cultivars) of finnochio fennel have been developed to suit different locations and climates. Varieties also differ in bulb shape, size, firmness, and time to maturity. In 1990, Indiana researchers conducted a variety trial with 16 fennel cultivars. Plants were spaced in single rows 3 feet apart. The researchers found that the Zefa fino variety had the highest yield of bulbs (452 lb per 1,000 square feet), the largest bulb circumference (8.1 inches), and the latest flowering dates. No other cultivar even came close in terms of yield or lateness of flowering.

The fennel plant will grow 2 to 5 feet tall. It is grown either for the seed or for the bulblike base. (Photo: Vince Rubatzky)

Market Information

Current production and yield. India, the People's Republic of China, Egypt, Argentina, Indonesia, and Pakistan make up the world's main fennel production areas. In California, fennel is grown in Monterey, San Luis Obispo, Santa Barbara, and Santa Cruz counties.

Use. The fresh leaves and tender leaf stems of sweet fennel are used as a boiled vegetable, and are sometimes served raw in salads or with other vegetables. The flavor of fennel is like that of anise or licorice. Unlike dill, fennel foliage cannot be dried. It is too sappy for successful drying, and by the time it has dried most of its flavor has been lost. Treat the flower heads as you would dill: sun-dry them for a few days, shake the seeds loose, and store the seeds in airtight containers.

Fennel is a pungent herb, slightly similar to dill when fresh, but not at all similar when dry. Varieties vary in their suitability for cooking — the best is Florentine sweet fennel, if the bulb is needed. For foliage and seeds, the best fennel is found in the Mediterranean. Although fennel grows wild in California, the climate is probably a little too moist here. Fennel is very good for digestion, and that is why it is often used with pork and in sausages in Italy. It is also the favorite herb with suckling pig and wild boar, dishes dating back to Roman times when fennel was used a great deal. For centuries, fennel has been used as a flavoring for drinks. One example, sack, is mentioned in William Shakespeare's *The Tempest.*

Use fresh leaves, tender stems, and young flower heads raw in salads, or use the bulb in stews and with chicken, fish, pork, beef, or cheese. Complementary wines are dry whites or Italian reds, depending on the dish.

Culture

Climatic requirements. Cool weather is best for the growth of fennel. Temperatures of 70° to 75°F are preferred. Seeds germinate best at soil temperatures of 61° to 64°F.

The plant has a tendency to bolt (flower prematurely) in warm summer weather. If you are growing fennel for the bulb, select a variety that will grow for a long time before bolting in order to produce more well-developed bulbs.

Propagation and care. Propagate by seed in spring or autumn. Plant in medium to light well-drained soil in a sunny location. Fennel thrives on well-drained loam soil. It grows well in mild temperate climates. For the seeding types of fennel, drill seed ½ inch deep in rows 1½ to 3 feet apart. Thin to 6 to 12 inches' spacing in the row when the plants are 3 to 4 inches tall.

Fennel is generally grown as an annual or biennial. Seed yields are low the first year, but increase the second year. Typically, the seeds mature in the fall of the second season. The umbels do not mature evenly, so multiple harvests will maximize yield. When the fruiting umbels turn brown, they are ready for harvest and should be cut promptly to prevent shattering. Machine harvesting, drying, and threshing may be possible, but will probably cause too much shattering loss and mixing of mature and immature seed.

Bulb anise should be thinned to 10 to 14 inches' spacing when grown on 40-inch beds with double rows 12 inches apart. Anise prefers acid soils, pH 5 to 6.8. The edible leaf base becomes fully mature in about 80 days but it is edible as soon as it begins to fatten.

Fennel prefers frequent irrigation for rapid growth. Moisture stress causes the basal stalk to split.

Pest and weed problems. Fennel competes poorly with weeds. Aphids can cause severe damage.

Harvest and postharvest handling. Harvesting, cleaning, trimming, and packing of the bulb (or finnochio fennel) is done by hand. Care must be taken to ensure a high-quality fresh pack. The foliage should be dark green and fresh in appearance, and the stalk and bulb a lighter greenish white. Harvest just before flowering. According to Morales, commercial growers remove the top foliage by cutting or "topping." The rest of the plant, including the bulbs, roots, and some foliage, is transferred from the field to a storage location where the roots are removed. Finally, the fennel bulb, with some foliage, is cleaned of soil, washed, and stored.

After harvest store the bulbs at 32° to 36°F. At the retail level, the bulbs are sold individually and may be wrapped in plastic. Market demand and price are based on bulb size, shape, and visual appearance. The bulb should be firm and free from insects and discoloration.

When the fennel is to be sold as an herb, use only the highest-quality plant material for the fresh market. Some of the detailed postharvest handling information provided for basil in this publication also applies to fennel and other herbs.

Sources

Seed

W. Atlee Burpee & Co., 300 Park Avenue, Warminster, PA 18974

Johnny's Selected Seeds, Foss Hill Road, Albion, ME 04910

Le Jardin du Gourmet, P.O. Box 75, St. Johnsbury Center, VT 05863

Park Seed Co., Cokesbury Road, Greenwood, SC 29647–0001

Seeds Blüm, Idaho City Stage, Boise, ID 83706

Shepherd's Garden Seeds, Shipping Office, 30 Irene Street, Torrington, CT 06790

Taylor's Herb Gardens, 1525 Lone Oak Road, Vista, CA 92084

More information

California Agricultural Statistics Service. 1987, 1988, 1989. *County Agricultural Commissioner data. 1987, 1988, and 1990 annual reports.* California Department of Food and Agriculture, Sacramento, CA.

Kowalchik, Claire, and William Hylton, eds. 1987. *Rodale's illustrated encyclopedia of herbs.* Rodale Press, Emmaus, PA.

Morales, Mario, Denys Charles, and James Simon. 1991. Cultivation of finnochio fennel. *The Herb, Spice, and Medicinal Plant Digest.* 9(1).

Simon, James. 1990. Essential oils and culinary herbs. In J. Janick and J. E. Simon, eds., *Advances in new crops.* Timber Press, Portland, OR.

Simon, James, Alena Chadwick, and Lyle Craker. 1984. *Herbs: An indexed bibliography 1971–1980.* Archon Books, Hamden, CT.

Stephens, James. *Minor vegetables.* 1988. Cooperative Extension Bulletin SP-40, University of Florida, Gainesville, FL.

Prepared by Curt Robinson and Claudia Myers.

Fresh Figs

Ficus carica is a member of the Moraceae (mulberry) family.

The fig is related to the tropical banyan and rubber trees, osage orange, mulberry, and che. The tree is deciduous, but is damaged by prolonged temperatures below 20°F. Its trunk has a swollen appearance, with heavy, succulent branches. Milky sap is present in stems, leaves, and unripe fruits. The sap's major constituent, Ficin, is harmful and irritating to skin and mouth.

Market Information

California produces the entire domestic U.S. crop of fresh figs. Because they are scarce and highly valued, figs are among the most profitable small farm crops for shipment to distant markets.

Use. Successful growers have long-term commitments from restaurants and other regular vendors, but must have easy access to an air freight service and must be able to produce a prime-quality pack over several months of summer and fall.

Figs are a demanding fresh fruit crop, and are grown almost exclusively as a small farm commodity. They are best suited to diversified farms where odd hours of labor are available for alternate-day pickings during summer.

Both purple- and green-fruited cultivars sell well. Unlike fresh figs, dried figs are produced in large-scale plantings where machine harvesting is most efficient.

Culture

Propagation and care. Plantings of one-half acre or less may use commercial nursery trees. For larger plantings use a commercial source for mother trees and take cuttings from these in following years. Do not collect cuttings from old, abandoned trees since they often carry fig mosaic virus. Stick foot-long cuttings into a prepared outdoor nursery bed. Transplant the next winter, with a tree spacing of at least 15 by 20 feet. Trim trees to 2 feet above the ground, and prune the following three winters to establish a stout scaffold and encourage branching.

Long outside and short inside pruning at waist height is best for hand-picked figs. This practice results in short trees that resemble head-trained

Green Ischia, a good fig for all California climates. (Photo: C. Todd Kennedy)

Fruit color and texture change as figs ripen. (Photo: Marita Cantwell de Trejo)

grapes, but it also delays initial fruit production by 3 to 4 years. Trees pruned in this manner require painting to prevent sunburn injury to the trunk and scaffold.

Irrigation is a must for good-quality crops, but fertilization is seldom necessary after initial planting. Overfertilizing leads to excessive extension growth with long internodes, a condition that must be corrected over several seasons.

Pests and disease. Figs are susceptible to vinegar flies and dried-fruit beetles, which affect marketability and introduce rot-causing microorganisms.

Orchard sanitation is critical. Do not plant figs near other fruit crops (particularly citrus) where fallen fruits are common. Harvest on a regular schedule. Collect and destroy any fallen or leftover fruits. Spraying may be necessary near citrus, peach, or plum orchards. Granular insecticides (applied in winter) are usually ineffective where nearby orchards serve as reservoirs for reinfestation. Epidemic souring of the fruit can be controlled only by rigorous spraying and fruit stripping over at least one season.

Strip and destroy all immature fruits immediately after the first frost to prevent *Botrytis* infection. Gophers can also damage fig orchards.

Harvest and postharvest handling. A fig tree has two harvest seasons: the Breba harvest in early summer and the main harvest in August. Breba figs may reach twice the size of main-crop figs and are very profitable, but they aren't as plentiful as main-crop figs. Many noncommercial cultivars do not produce the Breba crop.

Common fresh-market cultivars do not require pollination in order to produce fruits. The early Breba crop is borne on previous-year's growth, and late-season figs are borne on current-year's growth. Dormant pruning will eliminate the following year's Breba crop and greatly stimulate vertical growth of the tree.

Fresh fig harvesting is labor-intensive. Fruits are tender and easily damaged by handling. Figs are fully ripe when the stem begins to relax and fruit hangs limp on the tree. For shipping, fruit must be picked immediately before this stage, when the skin yields to pressure or shows characteristic ripening cracks. Fruit must be picked directly into cell packs or baskets with no further handling. Use only single-layer flats, and hurry the fruit to refrigeration and to market. Bruising will reduce fresh figs to unsalable condition within a few minutes.

Cultivars. Choose a two-crop variety for fresh figs: Desert King, Mission, Green Ischia, or Kadota in coastal counties; Beall, Mission, Conadria, or Kadota in the Central Valley. White Genoa makes a profitable late crop (to December) in the north. Brown Turkey is the variety of choice in Southern California, ripening continuously from mid-June through September near Los Angeles. Brown Turkey can also produce a very late crop in the Imperial Valley.

Sources

Nursery stock

NOTE: Desert King, Kadota, Black Mission, Brown Turkey, and other varieties are available from most wholesale fruit tree growers, especially those with garden center accounts.

L. E. Cooke Co., 26333 Road 140, Visalia, CA 93277

Orange Country Nursery, 27291 Road 92, Visalia, CA 93277

More information

Federal-State Market News Service. 1989. *Los Angeles fresh fruit and vegetable wholesale market prices 1989.* California Department of Food and Agriculture Bureau of Market News and USDA Marketing Service.

Federal-State Market News Service. 1990. *San Francisco fresh fruit and vegetable wholesale market prices 1990.* California Department of Food and Agriculture Bureau of Market News and USDA Marketing Service.

Prepared by C. Todd Kennedy.

Garlic Chive, Chinese Chive, Gow Choy

Allium tuberosum is a member of the Amaryllidaceae (amaryllis) family.
Varieties are usually listed as Chinese Chive or Garlic Chive.

This perennial plant grows in a clump and spreads by means of well-developed rhizomes. The plant grows to 12 to 18 inches' height. Each poorly developed bulb has four to five long, thin, flat leaves that bend downward at the tip. Garlic chives are grown very little in the United States, but extensively in China and in Japan. The flowers, usually white but sometimes purple, are distinctively shaped, starlike with a flat top umbel, more like those of the common chive than like the round ball that typifies garlic, leek, or onion.

Other names. *Ndoh dah* (Hmong); *nira* (Japanese); *jiu tsai, kau tsai,* or *kui tsai* (Mandarin Chinese); *gow choy* (Cantonese Chinese); *gil choy* (Chinese).

Market Information

By blanching Chinese chives, you can increase their price to $3 of $4 per pound. Blanching makes the chives more aromatic. To blanch, use tents of dark paper or some other material to exclude light from the growing plants. You can alternate crops of green and blanched chives.

Use. The flat, grasslike leaves are used like common chives. The flowers are also commonly eaten. Both have a garlic flavor. The bulbs can also be harvested and used like garlic cloves. Chinese chives have long been used as an herbal medicine to aid recovery from fatigue.

Culture

Propagation and care. Chinese chives prefer cool climates or temperatures. Production is best in spring. Flowering comes in the summer, and most varieties go dormant in hot weather. Growth will resume in the fall and become slow in winter. Chinese chives can be started from seed, usually in late winter or early spring. Once started, proper culture can keep the same planting productive for 10 years or more. Frequent nitrogenous fertilization and frequent but light watering are important for continued growth.

Chinese chives (garlic chives) grow to heights of 12 to 18 inches. (Photo: Mike Murray)

Chinese chives at the Los Angeles wholesale market. The flat grasslike leaves have a garlic flavor. (Photo: Hunter Johnson)

Sources

Seed

American Takii Inc., 301 Natividad Road, Salinas, CA 93906

Johnny's Selected Seeds, Foss Hill Road, Albion, ME 04910

Nichols Garden Nursery, 1190 North Pacific Highway, Albany, OR 97321

Park Seed Co., Cokesbury Road, Greenwood, SC 29647–0001

Sakata Seeds, 18695 Serene Drive, Morgan Hill, CA 95037

Seeds Blüm, Idaho City Stage, Boise, ID 83706

Shepherd's Garden Seeds, Shipping Office, 30 Irene Street, Torrington, CT 06790

Sunrise Enterprises, P.O. Box 10058, Elmwood, CT 06110–0058

Taylor's Herb Gardens, 1525 Lone Oak Road, Vista, CA 92084

Tsang and Ma, P.O. Box 5644, Redwood City, CA 94063

More information

Brewster, James, and Haim Rabinowitch. 1989. *Onions and allied crops.* CRC Press, Inc., Boca Raton, FL.

Harrington, Geri. 1984. *Grow your own Chinese vegetables.* Garden Way Publishing, Pownal, VT.

Jones, Henry, and Louis Mann. 1963. *Onions and their allies.* Interscience Publishers Inc., New York, NY.

Rubatzky, Vincent, and Mas Yamaguchi. 1997. *World vegetables, 2d ed.* Chapman and Hall, New York, NY.

Prepared by Ron Voss and Claudia Myers.

Gooseberry

Ribes uva-crispa and *Ribes hirtellum* are members of the Saxifragaceae (saxifrage) family.

Most gooseberries are derived from two species: the European gooseberry (*Ribes uva-crispa*), native to the Caucausus Mountains and North Africa, and the American gooseberry (*R. hirtellum*), native to the northeastern and north central United States and Canada. The gooseberry bush has thorny, arching branches that give the plant a height and breadth of 3 to 5 feet. Flowers are self-fertile and are pollinated by wind and insects, but usually not by honey bees. The fruit may be green, white, or yellow, or various shades of red. Fruits of different cultivars range from pea-sized to the size of a chicken's egg. Fruit characteristics and flavors vary from sour pulp and tough skins to tender skins and aromatic, sweet pulp.

Market Information

Use. Gooseberries are picked slightly underripe for cooking in jams, pies, and a classic dish, "gooseberry fool," made by folding cream into the stewed fruit. Fully ripe fruits of Poorman, Whitesmith, Hinnomakis Yellow, Achilles, and Whinham's Industry cultivars are excellent eaten fresh. Gooseberries are popular among northern Europeans, but little known in America. Before 1966, U.S. federal law banned the gooseberry plant as a cause of white pine blister rust. Since the ban was lifted, interest is growing in the United States.

Culture

Climatic requirements. Gooseberries tolerate temperatures to –30°F or lower, depending on the cultivar. They do not tolerate extremely hot summers, particularly if they are very dry. The bushes thrive in cool, well-drained, fertile soils. In warm regions, bushes perform better in heavier soils that retain moisture and keep cooler. An organic mulch helps protect the shallow roots. The gooseberry bush will thrive in full sun or in partial shade. In warmer summer areas, plant them in partial shade or on a north-facing slope.

Propagation and care. American cultivars are easier than European cultivars to propagate from cuttings. Hardwood cuttings about 1 foot long (not including tip growth) can be taken in the fall, preferably before all leaves have fallen. Tip and mound layering are more reliable methods of prop-

These gooseberries are ripe and ready for harvest. (Photo: Bernadine Strik)

agation, but a single bush furnishes fewer tip layers than cuttings, and mound layering sacrifices the fruit crop for the whole growing season. Either way, roots form where the stems are in the soil, and the small plants will be ready for transplanting by the first or second fall, depending on the cultivar.

Plant bushes 4 to 6 feet apart depending on cultivar vigor and soil. Do not plant bushes close to one another, or the arching branches of adjacent bushes will make pruning and picking difficult. Set bare-root plants in the ground in the fall or as early as possible in late winter or early spring.

Gooseberry bushes can be grown in either of two ways: on a permanent, short "leg," which is a trunk about 6 inches long, or as a "stool," where the bush is continually renewed with new shoots arising at or near ground level. Leg bushes grow large fruit that are easier to pick and are more accessible for pest control, but significant damage to the single leg can mean the loss of the entire bush. Stool bushes live longer and bear greater numbers of fruit than leg bushes, but their fruit are smaller.

To grow the gooseberry bush on a short leg, start with a cutting from which you have removed all but the uppermost four or five buds. You may instead want to leave all buds in place to encourage rooting, and then pull (but not cut) the lower buds off when you lift the young plants for transplanting.

The winter after the leg plant's first growing season, remove all but three or four vigorous branches pointing upward and outward. Head these primary branches back to 6 inches to stiffen them and induce further branching. Next winter, head the new secondary branches similarly—there should be a half-dozen or so. These will be the plant's permanent leaders, and should be pruned each subsequent winter by one-quarter of their new growth. Fruiting and age will slow the leaders' growth, so all they eventually will need will be a light tipping or nothing at all. Snap off any new branches that form along or below the initial 6-inch leg.

Off these leaders, lateral branches will grow that can be left to bear fruit along their whole length or shortened to make fewer, larger fruits. At the very least, cut away any laterals that are crossing, drooping, or otherwise misplaced. Another approach is to shorten all laterals to about 5 inches in early July, and then to cut them further back to about 2 inches during winter. This close pruning has the benefit of cutting away some mildewed branch tips and keeping the bush open to air, sun, and sprays. Such plants also make fruit picking easier.

To grow the bush as a stooled plant, you need not remove any buds from the initial cutting. The winter after the plant's first season in the ground, cut away all but about four of the previous season's shoots. Do the same thing the second winter so the bush will have four 1-year-old and four 2-year-old shoots. After the third winter's pruning, the bush will have four each of 1-, 2-, and 3-year-old shoots.

In the fourth and subsequent winters, cut away all 4-year-old shoots and all but about four of the most recent year's new shoots that grew up from ground level. The bush will then continue to have four each of 1-, 2-, and 3-year-old shoots. Except for lanky shoots that need shortening, all pruning of a stooled plants is done by cutting away branches at ground level. An excess number of canes may reduce fruit size and quality and may increase the plant's susceptibility to powdery mildew.

A gooseberry bush has a moderate need for nitrogen and a high potassium requirement. Avoid potassium deficiency with an annual dressing of ½ ounce of actual potassium per square yard. When liming the soil, use dolomitic limestone to provide the magnesium that the plants also need.

Pests. American gooseberry mildew can ruin gooseberries overnight under certain weather conditions. Fungicide sprays and correct fertilization, especially with nitrogen and potassium, can limit the disease. Leafspot can be controlled with orchard sanitation and sprays. Cultivars vary in their susceptibility to leafspot and mildew. The imported currantworm and fruitworm can be controlled with insecticides early in the season.

Sources

Plants

Alexander Eppler Ltd., P.O. Box 16513, Seattle, WA 98116-0513

Edible Landscaping Nursery, Rte. 2, Box 77, Afton, VA 22920

International Ribes Association, c/o Anderson Valley Agricultural Institute, P.O. Box 130, Boonville, CA 95415

Plumtree Nursery, 387 Springtown Rd., New Paltz, NY 12561

Raintree Nursery, 391 Butts Rd., Morton, WA 98356

Southmeadow Fruit Gardens, Lakeside, MI 49116

Tolowa Nursery, 360 Stephen Way, Williams, OR 97544

Whitman Farms Nursery, 1420 Beaumont NW, Salem, OR 97304

More information

Antonelli, A., et al. *Small Fruit Pests — Biology, Diagnosis, and Management.* 1988. Publication EB 1388. Washington State University Agricultural Communications, Pullman, WA.

Baker, Harry. 1986. *The fruit garden displayed.* Cassell Ltd., The Royal Horticultural Society, London.

Darrow, G., and S. Detwiler. 1924. *Currants and gooseberries: Their culture and relation to white-pine blister rust.* Farmer's Bulletin No. 1398. USDA, Washington, DC.

Galletta, G., and D. Himelrick, eds. 1990. *Small fruit crop management.* Prentice Hall, West Nyack, NY.

Ourecky, D. K. 1977. *Blackberries, currants, and gooseberries.* Cooperative Extension Publication IB 97. Cornell University, Ithaca, NY.

Reich, Lee. 1991. *Uncommon fruits worthy of attention: A gardener's guide.* Addison-Wesley, Reading, MA.

Prepared by Lee Reich.

Japanese Bunching Onion, Welsh Onion, Multiplier Onion

Allium fistulosum is a member of the Amaryllidaceae (amaryllis) family.
In China, it is called Cong; in Japan, Negi.

Varieties of Japanese bunching onion include White Spear, Evergreen, Kujo (Kujyo) Multistalk, Ishikura Long, He-shi-ko, Nebuka, Kincho, Red Beard, Tsukuba (a heat-resistant variety grown in spring for summer use), Multi-Stalk, and Tokyo Long White. In Japan, thick-bladed types are also grown. Among the popular varieties are Kaga, Shimonita, Senju, and Kuronobori. Hybrids with the common onion (*A. cepa*) include White Knight and Beltsville Bunching.

The Japanese bunching onion is a garden variety of perennial green onion that produces tillers from its stem. Although a perennial, it is typically grown as an annual. This onion does not form a bulb. It grows in clumps with several tillers bunched together. The stalks are silvery white and about ¾ inch in diameter. Depending on the variety, they will grow 6 to 24 inches long.

This is one of the few *Allium* species that can cross-pollinate with the common onion. As a result, many of the commonly grown green bunching onions are *Allium cepa* × *A. fistulosum* crosses. The non-bulbing and disease-resistant characteristics of *A. fistulosum* are combined with the great vigor and small leaf blades of *A. cepa*.

Other names. *Ndoh trah* (Hmong); *nebuka* (Japanese); *chung* (Chinese); *sibuyas* (Filipino); *hanh-ta* (Vietnamese).

Market Information

Giant Japanese bunching onions have a limited market but get a good price. Some markets would like to see more of the small varieties of bunching onions, such as purple varieties. Mexico has some economic advantage with this labor-intensive crop, so one way to compete is to grow a variety consumers prefer, such as the smaller varieties.

Use. The hollow tube-like leaves are used. The leaf blades and bases can also be used much like those of a common green bunching onion. Growers sometimes blanch the leaves by mounding soil around the lower leaf base to heights of 10 or more inches.

Japanese bunching onion grows 6 to 24 inches tall. (Photo: Hunter Johnson)

Culture

Climatic requirements. The Japanese bunching onion is generally a cool-season vegetable, but varieties have been adapted to a wide range of climates. Most varieties common to the United States are winter-hardy under most California conditions.

Propagation and care. Propagation is generally by seed, but the plants can be grown as overwintering perennials. Plant the seed ½ inch deep. Planting should be in the fall for spring harvest, early spring for summer harvest, or late spring for late summer harvest. Bolting can be a problem if you plant in the fall or winter. Japanese bunching onion needs well-drained soil.

You can thin the plants to a spacing of 3 to 4 inches since they will form bunches from a single seed, or you can plant them thick (½ to 1 inch apart) and transplant them when they reach 6 inches' height. The growing season lasts about 4 months (longer if planted in fall or winter).

Like other *Allium* species, Japanese bunching onion has a sparse, shallow root system. We recommend a complete fertilizer at planting time, followed later on by two or three additional nitrogenous fertilizer applications. Frequent, light irrigations are best. The plant competes poorly against weeds, so site selection and weed control practices are important.

Sources

Seed

American Takii Inc., 301 Natividad Road, Salinas, CA 93906

Johnny's Selected Seeds, Foss Hill Road, Albion, ME 04910

Nichols Garden Nursery, 1190 North Pacific Highway, Albany, OR 97321

Park Seed Co., Cokesbury Road, Greenwood, SC 29647–0001

Sakata Seeds, 18695 Serene Drive, Morgan Hill, CA 95037

Seeds Blüm, Idaho City Stage, Boise, ID 83706

Shepherd's Garden Seeds, Shipping Office, 30 Irene Street, Torrington, CT 06790.

Sunrise Enterprises, P.O. Box 10058, Elmwood, CT 06110–0058

Tsang and Ma, P.O. Box 5644, Redwood City, CA 94063

More information

Brewster, James, and Haim Rabinowitch. 1989. *Onions and allied crops.* CRC Press, Inc., Boca Raton, FL.

Harrington, Geri. 1984. *Grow your own Chinese vegetables.* Garden Way Publishing, Pownal, VT.

Jones, Henry, and Louis Mann. 1963. *Onions and their allies.* Interscience Publishers Inc., New York, NY.

Mansour, N. S. 1990. *Green bunching onions.* Vegetable Crops Recommendations. Oregon State University, Corvallis, OR.

Rubatzky, Vincent, and Mas Yamaguchi. 1997. *World vegetables, 2d ed.* Chapman and Hall, New York, NY.

Prepared by Ron Voss and Claudia Myers.

Jicama, Yam Bean

Pachyrrhizus erosus is a member of the Fabaceae (pea) family.

Jicama (pronounced "hee-<u>ca</u>-ma") is a tropical legume that produces an edible fleshy taproot. It is native to Mexico and northern Central America, and is widely cultivated there and in Southeast Asia. The plant is a vigorous spreading prostrate vine growing to several feet in diameter. Blue or white flowers and pods that look something like lima bean pods grow on fully developed plants. There are several species of jicama, but the one found in our markets is *Pachyrrhizus erosus.* The two cultivated forms are *jicama de agua* and *jicama de leche.* The latter has an elongated root and milky juice. The *agua* form has a top-shaped or oblate root and translucent juice, and is the only one marketed in the United States.

Market Information

Jicama is imported into the United States, mostly from Mexico. There are three important areas of jicama production in Mexico: Guanajuato and the central highland area, where the roots are harvested in autumn; the Morelos area, where they are harvested in winter; and Nayarit, where they are harvested in winter and spring. Despite recurrent interest in producing this crop in California, no successes have been documented. The known efforts to grow jicama have resulted in luxurious vine growth with prolific flowering and pod production, but low-quality fibrous taproots.

A long, warm growing season with relatively short days is required to initiate good development of the fleshy root. Research by Cotter and Gomez confirms this and suggests that the species may be sufficiently variable to allow selection for longer-day types. Since the temperature and day length conditions necessary for production of good-quality roots from currently available cultivars do not exist in the United States (except perhaps in south Florida), current cultivars probably cannot be grown successfully in this country. Any production of good roots in California has probably occurred when roots matured under unusually warm October and November conditions. This sort of weather is rare even in Southern California.

Use. Jicama is most commonly eaten fresh. After the fibrous brown outer tissue of the root is peeled

When mature, the jicama plant displays either blue or white flowers. (Photo: Hunter Johnson)

Harvested jicama roots. (Photo: Hunter Johnson)

away, the crisp white flesh can be sliced, diced, or cut into strips for use as a garnish, in salads, or with dips. It is frequently served as a snack, sprinkled with lime or lemon juice and a dash of chili powder. Jicama remains crisp after boiling and serves as a textural substitute for water chestnuts. Jicama is similar in food value to white potatoes, but with slightly fewer calories. In tropical production areas, the immature pods are sometimes cooked and eaten, but mature pods contain rotenone, and are said to be toxic. At one time growers considered commercial cultivation of jicama as a means of producing this insecticide.

Nutrition. Jicama has 20 mg of Vitamin C, 1.4 g of protein, 9 g of carbohydrates, and 15 mg of calcium in each 100 g edible, raw portion.

Culture

Jicama is propagated from seed. The seeds are squarish, brown or tan, and have the general characteristics of other bean seeds. Sandy loam soil with good drainage is the best choice for production of smooth roots. Rows should be 2 to 3 feet apart with plants every 8 to 10 inches in the row. Information on fertilizer requirements is limited, but one source suggests 1,500 pounds per acre of 6-6-12. In the tropics, development of marketable roots takes 3 to 6 months, depending on temperatures at the site. The literature on jicama indicates that for best root production growers should remove flowers at an early stage. Flower removal is said to cause the root to expand in diameter. Yields are in the range of 5 to 7 tons per acre.

Harvest and postharvest practices. Like potatoes, jicama may be harvested at any time during root development, but it is generally allowed to reach full size. Immature roots have a very tender skin. In Mexico, the mature roots are lifted out of the ground by hand or with a modified plow, selected for uniform shape and freedom from defect, and placed in baskets that are then emptied into trucks or trailers that transport the produce in bulk under ambient conditions to market.

Much of the jicama destined for U.S. markets is handled through intermediaries in Tijuana. There, the jicama is unloaded, washed with water, selected, trimmed (the taproot and remaining stem are cut off), packed into crates of about 50 pounds each, and dipped in a solution of about 10 percent calcium hypochlorite to sanitize and whiten the root. The crates are drained, palletized, and transported again under ambient conditions to U.S. wholesale markets. There the produce may be reselected and repacked and shipped anywhere in the United States in mixed-load shipments. In Mexico, jicama may be stored in the ground for as long as three months.

Store harvested jicama in a cool, dry area. Too much moisture will cause mold. Common postharvest problems include sprouting in storage, decay, and dehydration. USDA recommends storage at 55° to 65°F and 65 to 70% relative humidity, with an approximate storage life of 1 to 2 months. Research at UC Davis has concluded that jicama is sensitive to chilling, and that storage below 54.5°F causes major problems. Chilling injury symptoms include external decay and internal discoloration. Development depends upon both the temperature and the length of storage. For example, at the end of 2 weeks' storage at 50°F, the roots appear capable of recovery. At 3 weeks, however, the roots are permanently and seriously damaged. At 32°F, serious chilling damage occurs within 1 week. At 40°F, damage occurs within 1 or 2 weeks.

Sources

Seed

Gurney's Seed and Nursery Co., Yankton SD 57079

Hastings, P.O. Box 115535, Atlanta, GA 30310

J. L. Hudson Seedsman, P.O. Box 1058, Redwood City, CA 94064

Lockhart Seeds Inc., P.O. Box 1361, Stockton, CA 95205

Nichols Garden Nursery, 1190 N. Pacific Highway, Albany, OR 97321

Redwood City Seed Co., P.O. Box 361, Redwood City, CA 94064

Sunrise Enterprises, P.O. Box 10058, Elmwood, CT 06110-0058

Tsang and Ma International, P.O. Box 5644, Redwood City, CA 94063

More information

Alvarenga, A. A., and I. F. M. Valio. 1989. Influence of temperature and photoperiod on flowering and tuberous root formation of *Pachyrrhizus tuberosus*. Ann. Botany 64:411–14.

Cotter, D. J., and R. E. Gomez. 1979. Daylength effect on root development of jicama (*Pachyrrhizus erosus* Urban). Hort. Sci. 14(6):733–34.

Hansberry, R., R. T. Clausen, and L. B. Norton. 1947. Variations in the chemical composition and insecticidal properties of the yam bean. J. Agric. Res. 74:55–64.

Herklots, G. A. C. 1972. *Vegetables in South-East Asia*. George Allen and Unwin Ltd., London.

Lynd, J. Q., and A. A. C. Purcino. 1987. Effects of soil fertility on growth, tuber yield, modulation, and nitrogen fixation of yam bean (*Pachyrrhizus erosus* (L.) Urban) grown on a typic eutrustox. J. Plant Nutrition 10(5):485–500.

Miller, C. D., and B. Branthoover. 1957. Nutritive values of some Hawaii foods." *Hawaii Agric. Exp. Sta. Circ.* 52:16.

Orozco, Warner, Marita Cantwell, and Luis Hernandez. 1990. *Postharvest studies on jicama (Pachyrrhizus erosus). I. Quality and physiological changes of roots stored at different temperatures.* Progress Report. Department of Vegetable Crops, Mann Laboratory, University of Calivornia, Davis, CA.

Paull, R. E., and N. J. Chen. 1988. Compositional changes in yam bean during storage. *HortScience* 23(1):194–96.

Paull, R. E., N. J. Chen, and S. K. Fukuda. 1988. Planting dates related to tuberous root yield, vine length, and quality attributes of yam bean. *HortScience* 23(2):326–29.

Porterfield, W. M. 1939. The yam bean as a source of food in China. *New York Bot. Gard. J.* 40:107–108.

Porterfield, W. M. 1951. The principal Chinese vegetable foods and food plants of Chinatown markets. *Econ. Bot.* 5(1):12.

Schroeder, C. A. 1967. The jicama, a rootcrop from Mexico. *Proc. Trop. Reg., Am. Soc. Hort. Sci.* 11:65–71.

Sorensen, M. 1988. A taxonomic revision of the genus *Pachyrizus* (Fabaceae-Phaseoleae). *Nord. J. Bot.* 8(2):167–92.

USDA. 1987. *Tropical products transport handbook.* Agric. Handb. 668. USDA, Washington, DC.

USDA. 1984. *Vegetables and vegetable products.* Agric. Handb. 8–11. USDA, Washington, DC.

Prepared by Hunter Johnson, Jr., Warner Orozco, Marita Cantwell de Trejo, and Luis Hernandez.

Kiwano, African Horned Cucumber or Melon, Jelly Melon

Cucumis metuliferus is a member of the Cucurbitaceae (gourd) family.

The horned melon plant is a vine of African origin growing 5 to 10 feet long. The stem is angular, ridged, and hairy; internodes are 2 to 3 inches long. At each node, a 1- or 2-inch-long curling tendril forms along with two to four pale yellow male flowers, a leaf petiole, and occasionally a fruiting branch. The small, deeply cut, five-lobed leaves are similar to those of the watermelon. The fruits are oblong and 2 to 4 inches long when mature, and have long, sharp exterior spines. Immature fruits are light green. When they mature, they turn a bright orange on the outside, with green inner flesh.

Kiwano fruits are 2 to 4 inches long at maturity. (Photo: Hunter Johnson)

Market Information

The name "kiwano" is a registered trademark of Prinut Inc., which imports the horned melon from New Zealand. It's been grown in New Zealand since the mid 1980s. Frieda's Finest Produce Specialties of Los Angeles markets the kiwano in the United States. It is available year-round from New Zealand and California. Some marketers claim that California-grown horned melons are not sweet enough.

A New Zealand grower reported a yield of 200,000 10-pound trays of fruit from 50 acres, but an 18-acre planting at Los Baños, California, yielded only 12,000 trays.

Use. The spiny fruits have a bland citrus or banana-like flavor. The flesh of the ripe interior is lime green and jelly-like, with large seeds. The abundant, large, sticky seeds make the fruit difficult to use. The pulp can be strained for juice. The fruit turns bright orange when it is ripe. Horned melon is marketed as a garnish or for decorative purposes.

Culture

Climatic requirements. The plant is sensitive to cold, and will only grow during the warm seasons. Hot, dry conditions prevent powdery mildew.

Propagation and care. The plants grow very much like cucumbers. Be cautious when you grow them, since they have a "weedy" nature; robust and vigorous climbers, they can spread quickly. Fruits form in clusters; those closest to the center mature first. Wear gloves when cutting fruit from the vine, and avoid puncturing neighboring fruit with the spines. Harvest and packing are time- and labor-intensive.

Harvest and postharvest handling. The USDA storage recommendation is 50° to 60°F at 90% relative humidity, with an approximate storage life of 6 months. Do not stack the fruit, since the spines are likely to puncture other fruit. In New Zealand, the spines are made blunt with sandpaper or a file.

Sources

Seed
Seeds Blüm, Idaho City Stage, Boise, ID 83706

More information
Campbell, Dan. 1987. Hey spike. *Central Valley Rancher*, January 1987.

The Packer. 1990. *1990 produce availability and marketing guide*. Vance Publishing Corp., Overland Park, KS.

Shidler, Lisa. 1986. Kiwano sparks controversy. *The Packer,* July 19, 1986.

USDA. 1987. *Tropical products transport handbook*. Agric. Handb. 668. USDA, Washington, DC.

Prepared by Claudia Myers.

Kohlrabi, Stem Turnip

Brassica oleracea of the Gonylodes group is a member of the Brassicaceae (mustard) family. Varieties include White Vienna, White Danube, Green Vienna, Purple Vienna, Purple Danube, Earliest Erfurt, and Grand Duke.

Kohlrabi is a low biennial plant native to northern Europe. Cabbage-like leaves arise on long stems from the top and sides of its round, root-like stem. The stem's enlargement 1 to 3 inches above the ground accounts for kohlrabi's alternate name, "stem turnip." The leaves can grow as long as 10 inches. Two types of kohlrabi are available: one has green skin, the other has red or purple skin. The inner flesh is always white.

Kohlrabi (white Danube variety shown here) is grown for the enlarged stem. (Photo: Sakata Seed America, Inc.)

Market Information

Baby kohlrabi is tender and the greens are good tasting. This crop can be sold as a baby vegetable when it reaches a diameter of 1 to 2½ inches, or as an adult when it's about fist sized. The plant is hard to sell at intermediate sizes. Red or purple varieties are almost unknown as adults in many markets, but their leaves are better to eat than those of green kohlrabi.

Use. Before the kohlrabi stem is eaten, its peel is usually removed and its inner flesh is diced and boiled. Baby kohlrabi is steamed or otherwise cooked whole and served with butter. The greens are also valued at this stage; adult greens are too tough to eat. The swollen stem of a young kohlrabi may also be eaten raw; it is crunchy and can be served in salads or with dips. Cooks use the enlarged stem in essentially the same way as they use turnips. Use it when the tubers are 2 to 3 inches in diameter, before they become hard or bitter. Stress also causes the stems to become fibrous and woody.

Purple Danube kohlrabi. (Photo: Sakata Seed America Inc.)

Nutrition. Kohlrabi contains 20 IU of Vitamin A, 43 mg of Vitamin C, 1.7 g of protein, and 33 mg of calcium per 100 grams of raw, edible portion (about a cup).

Culture

Climatic requirements. Common varieties of kohlrabi may sprout unwanted seedstalks if grown for more than a week in cold temperatures of 50°F. The recommended temperature range is 65° to 77°F, with a 72°F optimum temperature. Cold spells with temperatures in the twenties will kill or stunt the plants.

Propagation and care. Propagation and time of planting are like those for cabbage; otherwise, kohlrabi is grown the same way as turnips. Plant spacing is about every 4 inches in rows 18 to 20 inches apart. Kohlrabi can also be planted with two rows on a 36-by-40-inch bed, one plant every 4 inches. Plant seeds approximately ½ inch deep. Kohlrabi matures in about 60 days when started from seeds, and in 40 days when started from transplants; however, transplanting may hurt development of the stems. Pests are the same as for cabbage.

Harvest and postharvest practices. Harvest the vegetable with clippers. Do not bunch, since that will bruise or break the leaves. Kohlrabi keeps well in transit if not packed too tightly. Kohlrabi is packed in bunches of two to five and placed in layers with the leaves toward the outside of the carton. The USDA storage recommendation is 32°F at 98 to 100% relative humidity, with an approximate storage life of 2 to 3 months for topped kohlrabi. The high humidity prevents shriveling and toughening. Kohlrabi with leaves has a storage life of only 2 weeks at 32°F.

Sources

Seed

NOTE: Kohlrabi seed is widely available.

More information

California Agricultural Statistics Service. 1989. *County Agricultural Commissioner data. 1989 annual report.* California Department of Food and Agriculture, Sacramento, CA.

Hardenburg, Robert E., Allen E. Watada, and Chien Yi Wang. 1990. *The commercial storage of fruits, vegetables, and florist and nursery stocks.* Agric. Handb. No. 66. USDA Agricultural Research Service, Washington, DC.

Larkcom, Joy. 1984. *The salad garden: Salads from seed to table.* Viking, New York, NY.

Liberty Hyde Bailey Hortorium. 1976. *Hortus third.* Cornell University/MacMillan Publishing Co., New York, NY.

Mansour, N. S. 1990. *Kohlrabi.* Vegetable Crops Recommendations. Oregon State University, Corvallis, OR.

The Packer. 1989. *1989 produce availability and merchandising guide.* Vance Publishing Corp., Overland Park, KS.

Rubatzky, Vincent, and Mas Yamaguchi. 1997. *World vegetables, 2d ed.* Chapman and Hall, New York, NY.

Stephens, James. *Minor vegetables.* 1988. Cooperative Extension Bulletin SP-40, University of Florida, Gainesville, FL.

USDA. 1975, 1984, 1989. *Composition of foods.* Agric. Handb. 8–11. USDA Agricultural Research Service, Washington, DC.

Prepared by Claudia Myers and Keith Mayberry.

Leek

Allium ampeloprasum, Porrum group, is a member of the Amaryllidaceae (amaryllis) family. Early maturing varieties include Tivi, King Richard, Gennevilliers, De Carentan, Titan, and Otina; late-maturing varieties include American Flag, Blue Leaf, Electra, Solaise, Alaska, Pinola, Laura, Nebraska, Carina, Conqueror, and London Flag.

Related to onions, chives, and garlic, the leek develops a broad, succulent stem rather than a large bulb like an onion. The mature stem is about 1 inch thick and 6 to 8 inches long from roots to the top of the neck; above the neck is a fanlike sheaf of flat, blue-green or yellow-green leaves that may grow another foot or two in length. Types vary in the length and thickness of the leaf shanks (pseudo-stem).

Market Information

Baby leek is a popular specialty item and can command a higher price per pound than larger leeks (i.e., greater than 1 inch diameter), but most leeks are sold at full size. Packing with top ice is highly desirable if you want to maintain the best quality.

Current production and yield. Most of the world's leeks are produced in Europe. California, New Jersey, Michigan, and Virginia are traditional leek-growing areas in the United States. Peak availability comes in late winter and in the spring, but leeks are available year-round.

The leaves of the leek plant are flat, in contrast with the round leaves of the onion. (Photo: Hunter Johnson)

Use. The thick, white leaf bases and slightly developed bulb are eaten as a cooked vegetable, in soups, or raw with or without leaves attached. The flavor is mild, sweet, delicate, and distinct from the other members of the onion family. The green leaves are edible, but have a pungent odor and an acrid taste. The leaves are used for flavoring in salads and cooked dishes.

Nutrition. Like all *alliums,* the leek is a good source of vitamins and minerals, especially Vitamins C, B_1, B_2, and B_6. A 3½-ounce portion provides about 30% of the recommended daily allotment for Vitamin C. Leeks also provide a good source of potassium. Excessively high amounts of nitrogen fertilizer can cause high nitrate concentrations on edible portions of the leaves and shanks.

The Nebraska leek variety at market. (Photo: Marita Cantwell de Trejo)

Culture

Climatic requirements. Leeks can withstand considerable exposure to temperatures below 32°F.

They grow best in cool to moderate climates. Because bolting (flowering) comes in the first spring after planting, late summer and early fall plantings should be avoided.

Propagation and care. Leek may be raised either by direct seeding or by transplanting approximately 3 months after greenhouse or bed planting. From the time of seeding, the plants may occupy the land for 8 to 15 months. For this reason, many growers prefer transplanting.

The fertilization program should be adjusted for the area, planting time, soil type, previous crop, and water management. Since the crop has shallow roots, water and fertility management will vary with soil type; sandier soils require more frequent irrigation and nitrogen application. Lighter soils help in harvesting and cleaning the crop; heavier soils may yield more leeks, but may also produce a lower-quality crop.

Spacing will influence the size of individual plants. The current trend is to grow long, thin shanks (1 inch diameter) with no noticeable bulbing and with as much white as possible. Wide spacing produces thick, short shanks. Such leeks may be fine for the processing crop, but they are less favored for fresh market.

When planning any planting, remember that leeks will bolt in late spring and early summer if they have been exposed to low temperatures. This limits summer production, since early fall plantings will bolt and spring plantings will be too small to yield well by late spring or summer.

Nursery planting for transplantation. In a greenhouse, sow seed approximately ⅜ to ½ inch apart in rows 12 to 14 inches apart. In outside, open beds, seed two to six rows to a bed, with beds 36 to 40 inches apart. The rows can be 4 to 14 inches apart on the bed. Plant the seed about ½ inch deep. No fertilizer is necessary in the nursery beds. Plant the seed at a rate of ½ to 1½ pounds per acre depending on seed size and desired population; one seed every ⅜ to ½ inch is optimal for nursery beds.

Nursery beds can be planted from October to April and transplanted 2 to 4 months later, when the plants are ⅜ to ½ inch in diameter and about 5 to 8 inches long. Trimming the roots and tops to make the transplants about 5 to 6 inches long is optional, and desirable only if the temperature is hot or if trimming makes the mechanical transplanting machine work better.

A spring (mid-May to June) harvest can be attained with October-to-December seeding; yields will be lowest at this time, but market price can be better. Summer harvest can be achieved with a December-to-February seeding; yields are usually highest at this time. Fall and winter harvests are usually achieved from direct seeding into fields during March, April, and May.

Transplanted crop. Set the transplants into beds similar to those used in the nursery at a spacing of 2 to 4 inches apart within the row and two rows per bed. Work a preplant broadcast application of 1,000 pounds of a 12–12–12 fertilizer or an equivalent complete fertilizer into the soil. Pull a shank down either side of the bed to make a narrow slot into which the plants can be hand planted every 4 to 6 inches. After planting, run a small double duckfoot shovel down the center of the bed or run shovels alongside the bed shoulders to move soil toward the plants. Strawberry equipment should work very well to firm-in the transplants. Irrigate as soon as planting is done. Several times during the growing season, apply side-dress nitrogen for optimum growth. You will want to supply an additional 100 pounds of nitrogen.

Direct seeding. Transplants might have the advantage of producing a longer white shank because they can set deeper into the soil. However, throwing the soil up against the plant every so often can help the plant to blanch and produce longer shanks. This produces better quality leeks, but also delays harvest. Direct seeding is less expensive, and determining whether to direct-seed or to transplant will depend on what kind of investment the grower wants to make, field availability, and the desired harvest time. Transplanting helps to minimize weed control problems.

The crop shouldn't be stressed for moisture; neither should it be overirrigated. Furrow, sprinkler, or drip irrigation can be used. The irrigation method is a matter of grower preference, uniformity in slope of the field, and water availability and cost.

Harvest and postharvest practices. A leek crop seeded in November and transplanted in February can be harvested in June; a crop seeded in March and transplanted in mid-June can be harvested as early as November, although the plants will not reach full size. A crop direct-seeded in April or May can be harvested the following spring. The market demand for leeks is usually greatest in the fall, winter, and early spring, and the bulk of the crop is harvested during this period. The largest plants are most susceptible to bolting in late spring or early summer. Because of high temperatures, leeks are not commonly harvested in California between June and October. Such harvest dates are possible in cool (e.g., coastal) climates.

Before harvest, plants are undercut with a blade attached to a tractor. The blade cuts about 1 to 2 inches beneath the plant base. Uniform undercutting without crop injury is easiest if the plants have been set uniformly at the transplant stage. The large, massive root system of the leek plant at harvest time and the fresh market's demand that 1 inch of clean roots be left on the plant combine to make harvesting difficult. After undercutting, each plant is lifted by hand and detached from the surrounding surface layer of soil with a knife. The soil at the base of the plant holds together in the extensive shallow roots like a mat. Although some soil can be shaken off in the field, the rest must be washed off with a pressurized stream of water. In muddy conditions, harvesting is a mess.

After the soil has been removed from the roots and an outside layer of the plant stripped clean by hand and washed, the topped or untopped plants are tied into bunches of 2 to 5 plants depending on size and packed into cartons. Some hand stripping is necessary to remove one or more sheaths of soiled or old leaves. Size grading would be a market advantage, but many growers do not go to this trouble. Presently, the market prefers some trimming of the tops. Harvesting is generally a weekly operation.

Storage recommendations are 32°F at 95 to 100% relative humidity, with an approximate storage life of 2 to 3 months. For shipping, join bunches at top, center, and bottom with ties.

Mature leeks ship in 20-pound (net weight), $4/5$-bushel crates holding 12 bundles each.

Pest and weed problems. Once a good canopy is developed, weed control should not be a problem. Before then, however, the leek is not very competitive. Direct seeding makes the problem worse. Both nursery and main crop areas should be rotated with non-alliums to guard against diseases and pests. Alliums should not be planted in the same soil or field more often than once every 4 or 5 years. Pink root can be a very serious disease. Thrips can also be a very serious insect problem.

Sources

Seed
NOTE: Leek seed is widely available.

More information
Brewster, James, and Haim Rabinowitch. 1989. *Onions and allied crops.* 1989. CRC Press, Inc., Boca Raton, FL.

California Agricultural Statistics Service. 1980–1988. *County Agricultural Commissioner data. 1980 through 1988 annual reports.* California Department of Food and Agriculture, Sacramento, CA.

Federal-State Market News Service. 1988. *Los Angeles fresh fruit and vegetable wholesale market prices 1988.* California Department of Food and Agriculture Bureau of Market News and USDA Marketing Service.

Jones, Henry, and Louis Mann. 1963. *Onions and their allies.* Interscience Publishers Inc., New York, NY.

Kline, Roger. 1987. Special vegetables. *Country Journal,* April 1987.

The Packer. 1989. *1989 produce availability and merchandising guide.* Vance Publishing Corp., Overland Park, KS.

Stephens, James. *Minor vegetables.* 1988. Cooperative Extension Bulletin SP-40, University of Florida, Gainesville, FL.

USDA. n.d. *Table of container net weights.* USDA Marketing Service, Washington, DC.

USDA. 1987. *Tropical products transport handbook.* USDA Agric. Handb. 668. Washington, DC.

Prepared by Ron Voss and Claudia Myers.

Lemongrass, Citronella Grass

Cymbopogon species belongs to the Poaceae (grass) family.
Lemongrass is *Cymbopogon citratus*. Citronella grass is *Cymbopogon nardus*.

Cymbopogon includes a number of aromatic perennial tropical grass species with mostly lemon-scented foliage. The gray-green stalks are 2 to 3 feet long and are as stiff as beach grass. The plant grows in dense clumps as big as 6 feet in height and diameter. *Cymbopogon nardus* is the source of commercial citronella oil. *Cymbopogon citratus* (lemongrass) is cultivated for the edible stem and for lemongrass oil.

Market Information

In 1988 the United States imported 74 metric tons of lemongrass oil, primarily from Guatemala and India. The value of this was estimated at $900,000.

Use. Inside the fibrous stem layers of lemongrass is a paler tubular core that resembles a firm scallion bulb. Cooks sliver this more tender part into various dishes, adding a pungent lemon flavor. It is also used in herbal teas and baked goods. Oil from lemongrass is used widely as a fragrance in perfumes and cosmetics, such as soaps and creams.

Culture

Climatic requirements. The best climate for lemongrass has a temperature range of 64° to 85°F and a high relative humidity (80 to 100%). In Stanislaus County, however, lemongrass is grown successfully at high temperatures (70° to 100°F) and low relative humidities (40 to 60%). The dry environmental conditions of the area favor the growth of the plant but not the growth of plant pathogens that attack the crop in its native tropical area. Lemongrass uses sunlight very efficiently, so it should be planted in areas exposed to the sun.

Cultural practices. Lemongrass grows best in well-drained sandy soils free of weeds and soilborne pathogens such as *Fusarium* and *Verticillium*. Since the plants rarely flower or set fruit, growers usually propagate plants using root cuttings or plant divisions. Stanislaus County growers place the plant cuttings in furrows 3 to 4 feet apart on beds 4 to 5 feet wide, depending on the variety and the duration of the season. Raised furrows will minimize crown root rot problems that could otherwise be induced by irrigation. Even though lemongrass is a perennial crop in the tropics,

Lemongrass grows in thick clumps up to 6 feet in height and width. (Photo: Charlotte Glenn)

Central Valley conditions will kill it off with first freeze. The crop is planted as soon as the danger of frost has passed. However, some Stanislaus County growers have successfully used plastic covers to protect the crop against cold weather. This permits harvest in March and April, when growers can get a premium price.

Fertilization. Lemongrass has nearly the same nutritional requirements as sweet corn. On average, growers in Stanislaus County applied 120 to 180 pounds of nitrogen and 140 to 180 pounds of phosphorus per acre, along with some potassium if needed. All of the phosphorus and potassium and about 15 to 20 pounds of nitrogen are placed 4 to 6 inches under the plant as starter fertilizer. The rest of the nitrogen is split and applied during the season as a sidedress or water run, preferably before or during an irrigation to improve plant uptake.

Irrigation. The lemongrass plant requires an average of 24 to 30 inches of irrigation water per year, depending on available soil moisture, soil type, and environmental conditions. Growers usually irrigate on an 8- to 10-day schedule. However, the shallow root system of lemongrass makes more frequent, lighter irrigations more beneficial. Moreover, the increased frequency of irrigations will increase the humidity, and that in turn will favor rapid plant growth.

Harvest. In Stanislaus County, lemongrass usually is harvested once a year, although elsewhere in the world it is harvested as many as four times per year. Stanislaus County growers harvest lemongrass by chopping the entire plant clump at the base. Tiller (stems) are then separated from the crown, cleaned, and bunched up for immediate sale. A bunch generally consists of six to eight stems, but sometimes four stems are enough. Consumers pay a premium for larger, thicker stems.

Sources

Root cuttings

Taylor's Herb Gardens, 1525 Lone Oak Road, Vista, CA 92084

Sunrise Enterprises, P.O. Box 10058, Elmwood, CT 05110–0058

More information

Liberty Hyde Bailey Hortorium. 1976. *Hortus third.* Cornell University/Macmillan Publishing Co., New York, NY.

Prasad, L. K., and S. R. Mukherji. 1980. Effect of nitrogen, phosphorus, and potassium on lemongrass." *Indian J. Agron.*

Schneider, Elizabeth. 1986. *Uncommon fruits and vegetables: A commonsense guide.* Harper and Row Publishers, New York, NY.

Simon, James, Alena Chadwick, and Lyle Craker. 1984. *Herbs: An indexed bibliography 1971–1980.* Archon Books, Hamdon, CT.

Prepared by Jesus Valencia and Claudia Myers.

Marjoram, Sweet Marjoram, Knot Marjoram

Origanum majorana and *Majorana hortensis* are members of the Lamiaceae (mint) family. Varieties include Sweet Marjoram and Creeping Golden Marjoram.

Marjoram is a tender perennial that grows about 1 foot tall and is treated as an annual in cold-winter areas. It has a dense, shallow root system and bushy habit. Marjoram leaves are pale gray-green, ¼ inch long, and grow on square stems. Tiny white, pink, or yellow flowers bloom in spikes in August and September. Fruits are very small, light brown nutlets.

Market Information

Use. Marjoram has medicinal, culinary, aromatic, cosmetic, ornamental, and craft uses. It has a spicy odor with a hint of balsam. Both leaves and flowers can be used fresh or dry as a food ingredient or a garnish. Cuisines of France, Italy, and Portugal make extensive use of marjoram.

As an aromatic, leaves are added to potpourris and sachets. Marjoram can be used as an ornamental plant in hanging baskets indoors in winter. Its large purple flower heads and furry, small leaves make it an attractive addition to winter bouquets. Fresh or dried sprigs can be added to herb wreaths.

Culture

Climatic requirements. Marjoram is hardy to zones 9 and 10. The plant grows best at temperatures ranging from 43° to 82°F.

Propagation and care. Marjoram thrives in full sun in fertile loam soil that is light, dry, and well drained. It grows in soil with pH values of 4.9 to 8.7. Seeds are small and slow to germinate. Seeds started indoors in mid-spring will germinate after 14 days at 60°F. You can sow the seeds directly in the garden after the soil has warmed, but keep the seedbed moist until seedlings have sprouted. Plants can also be propagated from cuttings, layering, or root divisions made in the late spring.

Set plants out when all danger of frost has passed, and space them 8 to 10 inches apart in rows 1 foot apart. Marjoram prefers conditions that are slightly more moist than its hardier relative oregano can tolerate. Mulch the plants to help retain soil moisture and keep weeds down. Cultivation may disturb marjoram's shallow root system. Water sparingly, but more than you would for oregano. Plants mature in 70 days.

Marjoram leaves near Hollister. (Photo: Hunter Johnson)

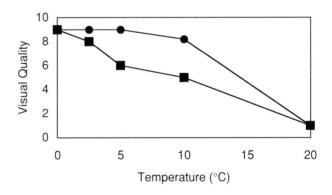

Effect of holding temperature on the quality of fresh marjoram after 1 (●) and 2 (■) weeks' storage in perforated polyethylene bags. Visual quality was assessed on a 5-point scale (1, 3, 5, 7, 9, where 1 = low quality and 9 = high quality). From Joyce, Reid, and Katz 1986.

Harvest. Harvest marjoram by clipping branches back to the bottom set of leaves. For a sweet and mild flavor, harvest the foliage before blooms begin to form. For a more pronounced flavor, harvest just before blooms open. Dry the harvested foliage away from sunlight to preserve both color and flavor. You can use a forced-air dehydrator at temperatures below 115°F. Once dried, discard the stems and store the crisp foliage in airtight and light-tight containers.

Mulched plants can be overwintered in milder areas. In colder areas, dig and divide roots early in the fall, bringing them inside for use in winter and for replanting outdoors in the spring. After 2 or 3 years, when plants are woody and less productive, divide and replant the root clumps.

Postharvest handling. Use only the highest-quality plant material for the fresh market. Some of the detailed postharvest handling information provided for basil also applies to marjoram and other herbs.

The successful marketing of fresh herbs requires careful postharvest handling. Temperature is the most important factor. The optimum postharvest temperature, 32°F, will allow a shelf life of 3 to 4 weeks; 41°F will allow a minimum shelf life of 2 to 3 weeks. Appropriate cooling methods for most herbs include cold rooms, forced-air cooling, and vacuum-cooling. Morning harvest minimizes the need for cooling.

Prevention of excess moisture loss is important. Most herbs respond well to high humidity: relative humidity in the packing area, cold rooms, and transport vehicles should be maintained above 95 percent where practical. You can also pack the fresh herbs in bags designed to minimize water loss. Maintain constant temperatures and reduce condensation inside the bags to prevent excess moisture and to avoid fungal or bacterial growth. Bags can be ventilated with perforations or fabricated from a polymer that is permeable to water vapor. Young herb tissue is susceptible to ethylene damage. This can be minimized by maintaining recommended temperatures.

If water is used during handling, chlorinated water can reduce the microbial load. To prevent physical injury to leaves, pack them in rigid clear plastic containers or pillow packs.

Sources

Seeds and plants

Abundant Life Seed Foundation, P.O. Box 772, Port Townsend, WA 98368

W. Atlee Burpee & Co., 300 Park Avenue, Warminster, PA 18974

Bountiful Gardens, 5798 Ridgewood Road, Willits, CA 95490

The Cook's Garden, P.O. Box 65, Londonderry, VT 05148

Gurney's Seed & Nursery Co., Yankton, SD 57079

Henry Field's Seed & Nursery Co., Shenandoah, IA 51602

Johnny's Selected Seeds, 299 Foss Hill Rd., Albion, ME 04910

Nichols Garden Nursery, 1190 North Pacific Hwy., Albany, OR 97321

Park Seed Co., Cokesbury Road, Greenwood, SC 29647-0001

Shepherd's Garden Seeds, 30 Irene Street, Torrington, CT 06790

Stokes Seeds Inc., Box 548, Buffalo, NY 14240

Territorial Seed Co., P.O. Box 157, Cottage Grove, OR 97424

More information

Cantwell, M., and M. Reid. 1986. Postharvest handling of fresh culinary herbs. *Perishables Handling* No. 60:2–4. Vegetable Crops Dept., University of California, Davis, CA.

Joyce, Daryl, Michael Reid, and Philip Katz. 1986. Postharvest handling of fresh culinary herbs. *Perishables Handling* No. 58:1–4. Vegetable Crops Dept., University of California, Davis, CA.

Kowalchik, Claire, et al., eds. 1987. *Rodale's illustrated encyclopedia of herbs.* Rodale Press, Emmaus, PA.

Newcomb, Duane, and Karen Newcomb. 1989. *The complete vegetable gardener's sourcebook.* Prentice Hall Press, West Nyack, NY.

Organic Gardening Magazine staff. 1978. *Organic gardening magazine's encyclopedia of organic gardening.* Rodale Press, Emmaus, PA.

Simon, James, Alena Chadwick, and Lyle Craker. 1984. *Herbs: An indexed bibliography 1971–1980.* Archon Books, Hamden, CT.

Prepared by Yvonne Savio and Curt Robinson.

Mung Bean

Vigna radiata is a member of the Fabaceae (pea) family.

The mung bean plant is 18 to 36 inches tall and produces a cluster of 2 to 8 slender, black pods 3 to 4 inches long. The slightly fuzzy pods contain very small seeds, which are green in the commercial varieties grown in the United States. Each pod may contain as many as 15 small oval seeds, depending on cultural conditions.

The mung bean is an ancient crop of Asia, where the seeds are available in many sizes, shapes, and colors. Domestic mung bean sprouts are produced from green-seeded cultivars. The mung bean's ancestors are annual plants with both short- and long-day cultivars. Most domestic cultivars are sensitive to temperature rather than daylength.

Other names. *Nong taao* or *pua sha* (Hmong); *moyashi-mame* (Japanese); *lu tou* (Mandarin Chinese); *look dou* (Cantonese Chinese); *balatung* (Tagalog); and *dau-xanh* (Vietnamese). Mung bean sprouts are known as *kolo taac* (Hmong); *moyashi* (Japanese); *yar tsai* (Mandarin Chinese); and *ngar choy* (Cantonese Chinese).

Market Information

The production of bean sprouts is the primary domestic use of mung beans. Sprout quality is an important factor in marketability. Whole, bright green seeds are likely to produce good sprouts. Damaging seeds during harvest, leaving mature seeds in the field too long before harvest (either on standing plants or in windrows), and molding associated with rain damage cause seed color to deteriorate. This will impair sprouting quality, rendering a seed lot unusable and unsalable. Early planting and timely harvest improve the chances of getting high-quality seed and sprouts. The marketability of mung beans depends on matching seed qualities to the particular needs of the sprouter, since sprouters differ in their equipment and techniques.

Current production and yield. Mung beans are produced in Oklahoma, Texas, and California. Other producers include Australia, India, Thailand, and other Asian countries. Sacramento Milling, Inc., in Ordbend is a major mung bean producer in the Sacramento Valley, producing seed under contract with growers.

The 18- to 36-inch-tall mung bean plant produces clusters of 3- to 4-inch-long pods. Each pod may contain as many as 15 small seeds. (Photo: Imo Fu)

Use. The principal domestic use of mung beans is the production of sprouts that are commonly used in Asian cooking. Seeds are germinated in the dark at 65° to 75°F in special germinating containers by wetting the seeds every 4 to 5 hours for 4 or 5 days. Commercial sprouters can expect an eightfold sprout yield (800 lb of sprouts for every 100 lb of seed). East Indian cuisine uses mung beans for *dal* (or *dahl*), a spicy paste made from the dry seed. Mung beans can also be shelled green and used like sweet peas, or the tender immature pods can be cooked whole. The mung bean is a staple legume in many diets around the world.

Culture

Propagation and care. Mung beans, like black-eye peas, are deep-rooted plants. Plants reach maturity in about 120 days. In the Sacramento Valley, growers should plant beans in May so the crop will be harvested before rain. Plant the seed 1 to 2 inches into moist, pre-irrigated ground, on 30-inch rows. Plant 3 or 4 plants per foot of row (15 to 30 pounds of seed per acre). Mung beans grow best with a few deep-penetrating irrigations. This program limits vegetative growth somewhat and forces photosynthetic partitioning to move into the seed. Generally, two or three post-emergence irrigations are adequate.

Harvest and postharvest practices. Mung bean pods are thin and brittle when dry, so shattering can be a problem. Harvest conditions and management often determine the success or failure of the crop. Use of a direct combine is the preferred harvest method, since it reduces seed shattering, which can be a significant factor in determining the quality of windrowed mung beans. Direct combining requires that the plant be defoliated and dried down in a timely manner. In the past, defoliants or desiccants have been used to achieve this end, but no registered chemicals are now available in the marketplace. Plants established in May will normally defoliate naturally and dry down in September.

In trials in Solano County, several fields were harvested with axial-flow combines using bean pickup heads. The growers reported that care should be taken to adjust the cylinder speed (<500 rpm preferred, 300 rpm optimum) and the screen plates to provide as clean a product as possible. Cylinder speed and screen plates appeared to be the main limiting factors; in general, the slower the better.

If you are going to cut and windrow, cut the plants when about half of the pods have turned black. There will still be many green pods, but most will continue to mature and dry. The thick stems retain moisture well even when cut. After cutting, allow the plants to dry undisturbed until harvest. If you cut and rake the plants, take care to handle them only once: additional handling will significantly increase shatter loss. Broken, cracked, or scarred seeds will not germinate properly, and so will result in lower sprout yields.

Pest and weed problems. Mung beans are susceptible to many of the same diseases and insect pests as black-eye peas. Mung beans secrete a sticky sweet substance that will attract a host of insects. Lygus, which can damage seed yield and quality, must be controlled during and after flowering. Black nightshade, hairy nightshade, yellow nutsedge, and summer annual grasses create harvest and quality problems if left uncontrolled. Aphids, cucumber beetles, mosquitoes, and several types of worms are also attracted, but they appear to cause little damage.

Sources

Seed

W. Atlee Burpee & Co., 300 Park Avenue, Warminster, PA 18974

Park Seed Co., Cokesbury Road, Greenwood, SC 29647–0001

Sacramento Valley Milling, P.O. Box 68, Ordbend, CA 95943

Sunrise Enterprises, P.O. Box 10058, Elmwood, CT 06110–0058

More information

Clement, Lawrence. 1986, 1987. *1986–87 Results of dry bean production studies in Solano County.* University of California Cooperative Extension, Solano County, CA.

Harrington, Geri. 1984. *Grow your own Chinese vegetables.* Garden Way Publishing, Pownal, VT.

Mackie, W. W. 1943. *The mung bean in California.* University of California Cooperative Extension, San Joaquin County, CA.

Murray, Mike. 1986. *1986 Colusa County mung bean nitrogen fertilizer field test results.* University of California Cooperative Extension, Colusa County, CA.

Park, H. G. 1978. *Suggested cultural practices for mung bean.* Guide 78–63. Asian Vegetable Research and Development Center, Taiwan.

Rubatzky, Vincent, and Mas Yamaguchi. 1997. *World vegetables, 2d ed.* Chapman and Hall, New York, NY.

Sandoval, Rick. 1986. Mung beans sprout hope for Solano diversity. *Ag Alert,* August 6, 1986.

Smith, Francis L. 1945. *Mung beans in California.* University of California, Agricultural Experiment Station, Berkeley, CA.

Stephens, James. *Minor vegetables.* 1988. Cooperative Extension Bulletin SP-40, University of Florida, Gainesville, FL.

Prepared by Carrie Young, Mike Murray, Lawrence Clement, and Claudia Myers.

Nappa Cabbage, Chinese Cabbage, Celery Cabbage, Pe-tsai

Brassica rapa **Pekinensis Group is a member of the Brassicaceae (mustard) family. Varieties include the Che-foo types: Wong Bok, Wintertime, Tropical Pride, Spring Giant, and Tokyo Giant; and the Chihili types: Michihili, Jade Pagoda, Market Pride, Shantung, and Shaho Tsai.**

There are two principal forms of Nappa cabbage. The Che-foo type (also known as Chiifu or Wong Bok, and sometimes Napa) develops a compact, often drum-shaped head slightly taller than it is wide. The head contains many broad, soft, light-green leaves. In the retail trade, nappa is usually understood to be this type. The other type, Chihili, forms a cylindrical head about three times as tall as it is wide. It has narrow, somewhat coarse leaves that are darker green than those of the Che-foo type.

Other names. Nappa cabbage is also known as *hakusai* (Japanese); *pai-tsai* (Mandarin Chinese); *won bok* (Cantonese Chinese); and *pechay* or *tsina* (Filipino).

Nappa cabbage packed for market. (Photo: Vince Rubatzky)

Market Information

Current production and yield. Nappa cabbage is available year-round from California and Hawaii. New York, Florida, New Jersey, Michigan, and Ohio are seasonal producers; they do not produce the crop in summer.

Culture

Climatic requirements. Nappa cabbage is a cool-season annual vegetable. It grows best with short days and moderate to cool temperatures (60° to 70°F mean temperatures).

Flowering (bolting) is favored by exposure to low temperatures (40° to 50°F) and long days. Such conditions during growth can cause premature bolting.

Temperatures greater than 80° to 85°F favor the development of tip burn, a nonpathogenic disorder. Compact and mature heads are more susceptible to tip burn.

Propagation and care. Nappa cabbage is relatively easy to grow. Plant it either by direct seeding or transplanting. Nappa cabbage plants require slightly more space per plant than regular cabbage. Use the same soil preparation, fertilization, and cultivation practices as you would for regular cabbage. Fertilizer requirements are similar to those for cauliflower. The crop will be ready 55 to 70 days after transplanting. Crop maturity can extend to 90 or 100 days with late varieties and less favorable (colder) growing temperatures. There is no exact age at which Nappa cabbage should be harvested, but heads should be firm and tight when ready. It may be eaten any time, once it has reached a usable size.

Harvest and postharvest practices. The USDA storage recommendation is 32°F at 95 to 100% relative humidity, with an approximate storage life of 2 to 3 months. Remove injured or diseased outer leaves before storage. Pack loosely and preferably upright in crates. Allow for air circulation in storage.

Sources

Seed

American Takii Inc., 301 Natividad Road, Salinas, CA 93906

Johnny's Selected Seeds, Foss Hill Road, Albion, ME 04910

Nichols Garden and Nursery, 1190 North Pacific Highway, Albany, OR 97321

Park Seed Co., Cokesbury Road, Greenwood, SC 29647–0001

Sakata Seeds, 18695 Serene Drive, Morgan Hill, CA 95037

Seeds Blüm, Idaho City Stage, Boise, ID 83706

Sunrise Enterprises, P.O. Box 10058, Elmwood, CT 06110–0058

Tsang and Ma, P.O. Box 5644, Redwood City, CA 94063

More information

Federal-State Market News Service. 1988. *Los Angeles fresh fruit and vegetable wholesale market prices 1988*. California Department of Food and Agriculture Bureau of Market News and USDA Marketing Service.

Federal-State Market News Service. 1987. *San Francisco fresh fruit and vegetable wholesale market prices 1987*. California Department of Food and Agriculture Bureau of Market News and USDA Marketing Service.

Harrington, Geri. 1984. *Grow your own Chinese vegetables*. Garden Way Publishing, Pownal, VT.

Kraus, James E. 1940. *Chinese cabbage varieties: Their classification, description, and culture in the central Great Plains*. Circ. 571. USDA, Washington, DC.

Mansour, N. S. 1990. *Chinese cabbage and leafy greens*. Vegetable Crops Recommendations. Oregon State University, Corvallis, OR.

The Packer. 1990. *1990 produce availability and marketing guide*. Vance Publishing Corp., Overland Park, KS.

Rubatzky, Vincent, and Mas Yamaguchi. 1997. *World vegetables, 2d ed*. Chapman and Hall, New York, NY.

Stephens, James. *Minor vegetables*. 1988. Cooperative Extension Bulletin SP-40, University of Florida, Gainesville, FL.

USDA. n.d. *Table of container net weights*. USDA Marketing Service, Washington, DC.

USDA. 1987. *Tropical products transport handbook*. Agric. Handb. 668. USDA, Washington, DC.

Prepared by Claudia Myers.

New Zealand Spinach

***Tetragonia tetragonioides* is a member of the Tetragoniaceae (carpetweed) family.**

The New Zealand spinach plant reaches a height of 1 to 2 feet and is much branched, spreading 2 to 3 feet across. New Zealand spinach is also known as *yeung poh tsoi* (Cantonese Chinese).

Market Information

Use. New Zealand spinach has a flavor very similar to but milder than that of common spinach. There is not a good market for it.

Culture

Climatic requirements. A heat-resistant, warm weather plant sensitive to frost, New Zealand spinach may find its niche as a summer green that can be grown in hot climates. It will tolerate saline soils.

Propagation and care. The large seeds should be soaked for a day before planting. Plants should be spaced 1 to 1½ feet apart in rows 2 to 3 feet apart. When the plant reaches a spread of 1 foot or so, you can use a knife to harvest the 2 or 3 inches of growth at the end of each branch, including tender shoots, tips, and leaves. These are cooked like spinach. New growth will arise along the cut branches, and you can then harvest their ends.

New Zealand spinach plants reach heights of 1 to 2 feet. (Photo: Hunter Johnson)

Commercial growers usually cut whole plants above the ground when they are small. New growth from the cut stem base then produces another crop. New Zealand spinach is bothered by few pests.

Sources

Seed

W. Atlee Burpee & Co., 300 Park Avenue, Warminster, PA 18974

Le Jardin du Gourmet, P.O. Box 75, St. Johnsbury Center, VT 05863

More information

Liberty Hyde Bailey Hortorium. 1976. *Hortus third.* Cornell University/Macmillan Publishing Co., New York, NY.

Stephens, James. *Minor vegetables.* 1988. Cooperative Extension Bulletin SP-40, University of Florida, Gainesville, FL.

Prepared by Claudia Myers (adapted from James Stephens's Minor Vegetables*).*

Okra, Gumbo

Abelmoschus esculentus **is a member of the Malvaceae (mallow) family.**

Okra varieties include Lee, Emerald (8-inch, smooth, dark green mature pods; 5-foot plant), Annie Oakley, Burgundy (red pods), Perkins Spineless (7-inch ridged green pods; 3-foot plant), Dwarf Green Long Pod (7- to 8-inch, slightly ridged green pods; 3-foot plant), Clemson Spineless (6-inch, moderately ridged green pods; 4- to 5-foot plant), and French Market.

The okra fruit (gumbo), a large, erect pod, is harvested immature. Okra is an herbaceous, shrub-like dicotyledonous annual plant with woody stems growing 3 to 6 feet tall. It has alternate broad leaves. The flower has five yellow petals and a purple area at the base. It is a tropical native of Africa and is related to cotton. Varieties are available with pods of differing lengths and colors (white, red, green, and purple).

Market Information

Green okra is the common type, but red or burgundy okra is considered a specialty crop.

Current production and yield. Top U.S. shipping states are California, Florida, and Texas. Most production is from June through November. Mexican imports are available year-round, but peak from May through October.

Use. Cooks use fresh okra pods to make gumbo, a well-known soup dish of the American South. The pods are also boiled as a vegetable dish, used in soups or stews, or fried, and are dried and used in the winter in certain parts of the country. Some people roast the seeds, grind them, and brew the resulting powder as a coffee substitute. The swelling gum (mucilaginous material) in okra is greater than in any other common vegetable, and may take getting used to.

Culture

Climatic requirements. Okra seeds are sensitive to cold, so they should not be grown until the ground has become warm. The plant grows best when minimum and maximum mean temperatures, respectively, are 65° and 95°F.

Okra growing in the Coachella Valley. (Photo: Hunter Johnson)

A red-podded okra variety. (Photo: Hunter Johnson)

Propagation and care. Sow the seed ½ to 1 inch deep, and then thin the plants to every 12 inches in the row. The rows are usually far enough apart (36 to 40 inches) to permit cultivation. Cotton equipment can be used for planting, fertilization, and cultivation. One ounce of seed will plant 100 feet of row (about 8 pounds per acre). Soaking the seed in room-temperature water for 24 hours will improve uniformity of germination. If the plants grow too tall, they may be cut back to about 2 feet and then fertilized with nitrogen for a new flush of growth. Okra does not tolerate wet, poorly drained, or acidic soils. In the desert valleys of southern California, growers can make early plantings from early February through March. In the San Joaquin Valley, an early planting can be made between the first of April and the middle of May.

Harvest and postharvest practices. Individual okra flowers open for only a day or so. The pods then develop quickly, and are ready for harvest 4 to 10 days after flowering, when the pods are 3 to 4 inches long. Because the pods develop so quickly, they must be harvested at least every other day. Typically, growers snap the okra pods from the plants by hand. Touching the plant causes skin irritation, so field workers should wear gloves and long-sleeve shirts when harvesting. Okra pods should be ready for first harvest about 70 to 80 days after planting, and will continue to bear for weeks.

To be graded US No. 1, pods must be fresh, tender, not badly misshapen, and free from decay and damage. Okra requires careful handling to prevent bruising (bruises blacken within a few hours), and should be packed into containers or baskets with net weights of 15, 18, or 30 pounds. Containers must be well ventilated because of okra's high respiration rate at warm temperatures. Okra should be cooled promptly upon harvest. Because it deteriorates rapidly, okra normally is stored only briefly.

The USDA storage recommendation is 45° to 50°F at 90 to 95% relative humidity, with an approximate storage life of 7 to 10 days for okra in good condition. Okra is sensitive to chilling — temperatures below 45°F cause discoloration, pitting, and decay.

See UDSA Agricultural Handbook No. 66 for more detailed information on postharvest handling of okra.

Pest and weed problems. Pests that occur occasionally on okra are, in their order of importance, cotton aphid, corn earworm, nematodes, Verticillium wilt, green stinkbug, harlequin bug, and pink bollworm. The pink bollworm does little damage to okra, but the crop serves as a bothersome host in quarantine programs. Because of the limited market for okra, there have been few programs to develop pesticide registrations and control recommendations for the pests. For current registrations, contact your local farm advisor or agricultural commissioner.

Cultural practices such as destroying or burying crop residues in the winter will help suppress bollworms. Crop rotation will control or suppress weeds, plant diseases, nematodes, and some insect pests.

Sources

Seed

NOTE: Okra seed is widely available.

More information

Aguiar, J. L., and K. S. Mayberry. 1997. *Okra Production in California*. Publication 7210. UC Division of Agriculture and Natural Resources, Oakland, CA.

California Agricultural Statistics Service. 1980–1989. *County Agricultural Commissioner data. 1980 through 1989 annual reports*. California Department of Food and Agriculture, Sacramento, CA.

Mansour, N. S. 1990. *Okra*. Vegetable Crops Recommendations. Oregon State University, Corvallis, OR.

Martin, Franklin W., and Ruth Ruberté. 1978. *Vegetables for the hot humid tropics. Part 2. Okra*. Science and Education Administration, USDA, New Orleans, LA.

The Packer. 1989. *1989 produce availability and marketing guide*. Vance Publishing Corp., Overland Park, KS.

Rubatzky, Vincent, and Mas Yamaguchi. 1997. *World vegetables, 2d ed*. Chapman and Hall, New York, NY.

Prepared by Claudia Myers.

Oregano, Winter Marjoram, Wild Marjoram, Pot Marjoram

Origanum spp. are members of the Lamiaceae (mint) family.
Besides oregano, this genus also includes marjoram (*Origanum majorana*).

Oreganos are aromatic, herbaceous perennials with erect, hairy stems that grow 1 to 3 feet tall. The oval green leaves are somewhat pointed and up to 2 inches long. Small tubular purple or white flowers appear from July through September. Each fruit contains four seedlike nutlets.

Of all varieties, Greek oregano (*O. vulgare* subsp. *hirtum* and *O. vulgare* Viride) produces the best culinary flavor. *Origanum vulgare* is grown for medicinal use; its pink flowers are used in floral arrangements. Other oreganos used in cooking are not *Origanum,* but are more heat tolerant in the garden. Mexican or Puerto Rican oregano (*Lippia graveolens*), a perennial, is a member of the verbena family grown in the south and southwest. *Poliomintha longiflora,* a native of Monterrey, Mexico, takes full sun or partial shade and tolerates freezing.

Oregano vulgare var. *aureum,* Creeping Golden Marjoram, is a fast-spreading plant that is useful in pathways and rock gardens. Trailing oregano grows well in hanging baskets or on rock walls.

Market Information

Use. Oregano has medicinal, culinary, cosmetic, craft, and companion planting uses. This herb has a strong, biting aroma and a sharp, piquant flavor. It is most commonly used as a fresh or dried culinary herb in Mediterranean and Latin American cuisines.

Culture

Climatic requirements. Oregano is hardy in plant climate zones 4 through 10, with temperatures ranging from 41° to 82°F. It thrives in full sun in well-drained, average-quality soil with a pH of 4.5 to 8.7.

Propagation and care. Propagate oregano from cuttings or by root division from plants. For culinary use, *Origanum heracleoticum* is more flavorful than *O. vulgare*. Take root divisions or cuttings in the spring. Space plants 12 to 15 inches apart in rows 18 inches wide after danger of frost has passed.

Flavor varies greatly when oregano is grown from seed. If you use seed, plant a lot so you will be able to select plants with the desired flavor. Seeds

Healthy Greek oregano plants. (Photo: Hunter Johnson)

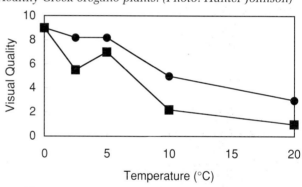

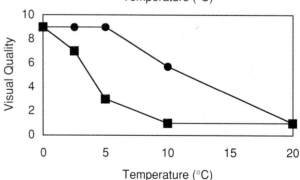

Effect of holding temperature on the quality of fresh Mexican oregano (top) and Greek oregano (above) after 1 (●) and 2 (■) weeks' storage in perforated polyethylene bags. Visual quality was assessed on a 5-point scale (1, 3, 5, 7, 9, where 1 = low quality and 9 = high quality). From Joyce, Reid, and Katz 1986.

germinate in 4 days at 70°F in the light; at cooler temperatures, they may take up to 14 days. If seeded outdoors in soil warmer than 45°F, cover seeds with cheesecloth to keep them from washing away.

Oregano is a vigorous grower and requires little attention. Plants mature in 45 days. Mulch the plants to keep foliage clean. Cutting a few sprigs of leaves when the plant is 6 inches high will encourage bushiness. Fertilize yearly with a balanced plant food. Little irrigation is needed after the plants are established. Divide plants every few years to prevent woodiness and declining productivity.

Pests and disease. Oregano is susceptible to root rot and to fungal diseases. It can be damaged by spider mite, aphid, and leaf miner infestations.

Harvest and postharvest practices. Use only the highest-quality plant material for the fresh market. Some of the detailed postharvest handling information provided for basil also applies to oregano.

To harvest oregano, trim all branches, leaving only the lowest set of leaves. The plant will leaf out again and send up new shoots within 2 weeks, providing another harvest. Foliage tastes sweeter when clipped before flowers begin to develop. Essential oils are greatest just before the plant blooms. To dry, hang the stems upside-down in a dark, dry place. Once dried, discard the stems and store the leaves in airtight, light-tight containers.

The successful marketing of fresh herbs requires careful postharvest handling. Temperature is the most important factor. The optimum postharvest temperature, 32°F, will allow a shelf life of 3 to 4 weeks; 41°F will allow a minimum shelf life of 2 to 3 weeks. Appropriate cooling methods for most herbs include cold rooms, forced-air cooling, and vacuum-cooling. Morning harvest minimizes the need for cooling.

Prevention of excess moisture loss is important. Most herbs respond well to high humidity: relative humidity in the packing area, cold rooms, and transport vehicles should be maintained above 95 percent where practical. You can also pack the fresh herbs in bags designed to minimize water loss. Maintain constant temperatures and reduce condensation inside the bags to prevent excess moisture and to avoid fungal or bacterial growth. Bags can be ventilated with perforations or fabricated from a polymer that is permeable to water vapor. Young herb tissue is susceptible to ethylene damage. This can be minimized by maintaining recommended temperatures.

If water is used during handling, chlorinated water can reduce the microbial load. To prevent physical injury to leaves, pack them in rigid clear plastic containers or pillow packs.

Sources

Seeds and plants

Abundant Life Seed Foundation, P.O. Box 772, Port Townsend, WA 98368

W. Atlee Burpee & Co., 300 Park Avenue, Warminster, PA 18974

Bountiful Gardens, 18001 Shaser Ranch Road, Willits, CA 95490

Gurney's Seed & Nursery Co., Yankton, SD 57079

Henry Field's Seed & Nursery Co., Shenandoah, IA 51602

Johnny's Selected Seeds, 299 Foss Hill Rd., Albion, ME 04910

Le Jardin du Gourmet, P.O. Box 75, St. Johnsbury Center, VT 05863

Nichols Garden Nursery, 1190 North Pacific Hwy., Albany, OR 97321

Park Seed Co., Cokesbury Road, Greenwood, SC 29647-0001

Shepherd's Garden Seeds, 30 Irene Street, Torrington, CT 06790

Territorial Seed Co., P.O. Box 157, Cottage Grove, OR 97424

More information

Cantwell, M., and M. Reid. 1986. Postharvest handling of fresh culinary herbs. *Perishables Handling* No. 60:2–4. Vegetable Crops Dept., University of California, Davis, CA.

Joyce, Daryl, Michael Reid, and Philip Katz. 1986. Postharvest handling of fresh culinary herbs. *Perishables Handling* No. 58:1–4. Vegetable Crops Dept., University of California, Davis, CA.

Kowalchik, Claire, et al., eds. 1987. *Rodale's illustrated encyclopedia of herbs.* Rodale Press, Emmaus, PA.

Newcomb, Duane, and Karen Newcomb. 1989. *The complete vegetable gardener's sourcebook.* Prentice Hall Press, West Nyack, NY.

Organic Gardening Magazine Staff. 1978. *Organic gardening magazine's encyclopedia of organic gardening.* Rodale Press, Emmaus, PA.

Simon, James, Alena Chadwick, and Lyle Craker. 1984. *Herbs: An indexed bibliography 1971–1980.* Archon Books, Hamden, CT.

Prepared by Yvonne Savio and Curt Robinson.

Parsnip

Pastinaca sativa is a member of the Apiaceae (parsley) family. Varieties include Avonresister, Gladiator, Tender and True, Hollow Crown, All American, and Harris Model.

The parsnip is related to the carrot, which it resembles at least in its root and habit of growth. Unlike carrots, parsnip roots are creamy white outside and white inside. The parsnip top resembles that of broadleaf parsley. The top grows to 3 feet tall and the root to 20 inches long, like a deep-rooted carrot.

Market Information

Current production and yield. In California, the County Agricultural Commissioners reports parsnips are grown in Imperial, Monterey, and San Luis Obispo counties. A yield of about 4 tons per acre is equivalent to 50 pounds per 100 feet of row.

Culture

Climatic requirements. Parsnips do best when they mature in cool weather. Parsnips take a long time to mature — up to 180 days. Cold weather (including a preharvest freeze) enhances the flavor. The seeds are difficult to germinate in hot weather, and roots maturing in hot weather are of poor quality.

Propagation and care. Start parsnips from seed as you would carrots. Normally, parsnips take 120 to 180 days from seeding to root harvest. Seed should be planted ½ inch apart and ½ inch deep in rows 12 inches apart, or planted in two rows on a 36- or 40-inch bed. About 3 to 5 pounds of seed will plant an acre. Parsnips take as long as 18 days to germinate. Thin the plants to every 3 or 4 inches in rows. Never allow the soil to get too dry or the roots will get tough, split, and lack flavor. Fertilize parsnips as you would carrots.

The parsnip is grown for its the carrotlike root, which grows as long as 20 inches. (Photo: Hunter Johnson)

Postharvest handling. The USDA storage recommendation is 32°F at 95 to 100% relative humidity, with an approximate storage life of 4 to 6 months.

Sources

Seed

Asgrow Seed Co., P.O. Box 5038, Salinas, CA 93915

W. Atlee Burpee & Company, 300 Park Avenue, Warminster, PA 18974

Ferry-Morse Seed Co., P.O. Box 4938, Modesto, CA 95352

Johnny's Selected Seeds, Foss Hill Road, Albion, ME 04910

Le Jardin du Gourmet, P.O. Box 75, St. Johnsbury Center, VT 05863

Nichols Garden Nursery, 1190 North Pacific Highway, Albany, OR 93721

Northrup King Co., P.O. Box 1825, Gilroy, CA 95021

Thompson and Morgan, P.O. Box 1308, Jackson, NJ 08527

More information

California Agricultural Statistics Service. 1980–1989. *County Agricultural Commissioner Data. 1980 through 1989 Annual Reports.* California Department of Food and Agriculture, Sacramento, CA.

Mansour, N. S. 1990. *Parsnips.* Vegetable Crops Recommendations. Oregon State University, Corvallis, OR.

USDA. 1987. *Tropical products transport handbook.* Agric. Handb. 668. USDA, Washington, DC.

Prepared by Keith Mayberry and Claudia Myers.

Prickly Pear Cactus

Opuntia species are members of the Cactaceae (cactus) family.

The genus *Opuntia* comprises the prickly pears, which include bunny ears and beaver-tail cacti. During a few weeks in late spring and early summer each pad produces several flowers. Depending on the variety, flowers may be yellow, orange, pink, or red. When the blooms fade, an edible fruit forms on many species.

The pads are fast-growing, flattened stems that vary in size with the variety. They can be 4 to 16 inches long, 3 to 9 inches wide, and up to ¾ inch thick. They have a smooth skin, may be elliptical to oblong, and are bright green to blue-gray. Most have numerous inch-long spines. They also have small stickers that easily penetrate a person's skin.

The fruits of all varieties are edible, but only a few varieties of fruit are palatable or sweet. Size, shape, and color vary. Fruit skin and flesh come in a rainbow of colors depending on variety. The sweetest varieties available in this country have dark reddish orange or purple skins and deep red-purple flesh.

Other names. Prickly pear is also called cactus pear and Indian fig, or *figadindi* in Italian. The pads are *cladodes* or *nopales* when whole and *nopalitos* when diced. The fruits are called prickly pears or *tunas*.

Market Information

Current production and yield. The prickly pear cactus is native to the United States, Mexico, and South America, but it grows well in other areas including Africa, Australia, and the Mediterranean region. The plant is particularly well adapted to arid zones, and it grows at elevations from sea level up to the high Andes, about 15,000 feet. In warm climates pads may be harvested as many as six times a year. Established plants may yield 20 to 40 half-pound pads at each harvest.

Use. The pads, fruit, and seeds can be eaten. The pads are eaten as a vegetable, served cooked or raw in a variety of dishes. The small young pads in the early spring are succulent, with a delicate flavor and the fewest spines. Fresh pads are full of water and should be bright green and firm. The pad is prepared by scraping the skin to remove spines, then peeling and preparing it according to a recipe.

Prickly pear pads and blossoms. (Photo: Yvonne Savio)

The fruits are typically eaten fresh, often sprinkled with lime juice after being refrigerated and peeled. Fruit flavor varies among varieties but is similar to that of strawberry, watermelon, honeydew melon, fig, banana, or citrus fruit. The fruit can also be cooked into jellies, desserts, beverages, and candies. The syrup is used in a potent alcoholic drink called *coloncha*. The pulp can be dried, ground into flour, and used for baking. Recipes and informative tips on preparation can be found in Joyce L. Tate's *Cactus Cookbook,* available from the Cactus and Succulent Society of America.

Prickly pear cactus is a very versatile plant. It can be used in landscaping, and the pads are a source of food and water for livestock and poultry. The sap is used medicinally like that of the aloe vera plant to soothe cuts and burns, and is also used in chewing gum and candles and as a stiffening agent for cotton cloth. The pads' strong fibers are woven into mats, baskets, fans, and fabrics.

Nutrition. The composition of the pads changes during development. At commercial size, they are approximately 92 percent water, 1.5 percent protein, 5 percent carbohydrate (including fiber), and 1.1 percent ash, and contain 13 mg Vitamin C per 100 gram portion.

The ripe fruit pulp is 85 percent water and 10 to 15 percent carbohydrates, and contains a significant amount of Vitamin C (25 to 30 mg per 100 gram portion).

Culture

Propagation and care. The prickly pear tolerates many soils, temperatures, and moisture levels. It grows best in sunny locations with well-drained sandy loam soil where it is protected from winter winds. A balanced fertilizer can be applied during the growing period. The plant is drought tolerant once established, but with good drainage it can tolerate a lot of water.

Opuntia species may be started from seed in a shady bed and kept moist until germination. Growth is slow. Flowers and fruits may not appear for 3 to 4 years. Propagation from pads is simpler and faster. Cut pads at least 6 months old from a growing cactus and allow them to form a callus. Then plant each pad upright, 1 inch deep into a mixture of soil and sand. Stake or anchor the pad to keep it upright until it roots, but do not water it in. Protect the potted pad from intense summer sun by orienting the slim side north to south. Roots will form in about a month. Water it then once, but allow it to dry between later waterings. You should wait several months before harvesting new pads so that future harvests will not be reduced. The second or third pad will bear flowers and fruit, but a pad taken from an older plant may flower and set fruit sooner than a pad taken from a younger, immature plant.

If you grow the prickly pear for its pads, feed the plant with a high-nitrogen fertilizer. If you prefer more flowers and fruit, give the plants a no-nitrogen fertilizer once a month, even through the winter. During this dormant period, the plants require bright light and just enough water to keep the pads from shriveling.

Depending on the variety, the cactus will bloom and set fruit from early spring through summer. Each pad can support a number of flowers, each yielding one fruit. You can allow several fruits on each pad and still get good-sized fruit.

Harvest and postharvest. Fruit ripens from early spring through late fall, depending on the variety. Varieties that are best eaten fresh ripen from September through November. The perfect stage of ripeness of each fruit lasts only about a week, and the maximum shelf life of a fruit is only 8 or 9 days.

The fruits are ripe enough to harvest when the glochids (small stickers) fall off. Twist the fruit from the pad rather than pulling it in order to avoid tearing. If the fruits are harvested unripe, they will not be as sweet. Remove the pads by carefully cutting them from their supporting pads. The best time to harvest pads is from mid-morning to mid-afternoon, when the acid content in the pads is at its lowest.

Depending upon the temperature, cladodes may take from 15 to 30 days to grow from bud to a commercially harvestable size, about 8 inches (20 cm) long. Good-quality cladodes are turgid and easy to break. They are shiny and green with the leaflets intact. Low-temperature storage (40°F) will increase cladode acidity, whereas storage at warmer temperatures (70°F) will result in a decrease. Cladodes are sensitive to chilling if held below 50°F. The most serious postharvest problems encountered in handling vegetable cladodes are water loss and rot development at the cut stem end.

Tunas are produced only on mature pads, require 110 to 120 days to develop, and are harvested by twisting the fruit carefully off the pad. The fruits are covered with small glochids (spines), which present the major difficulty in harvesting and handling. As fruit matures the glochids fall off, but the fruit also becomes more susceptible to physical injury so most are harvested commercially in an intermediate stage of ripeness. After harvest, the glochids are removed by sweeping the fruit on a grassy or straw-covered area or by sweeping them mechanically with rotating brushes. They are packed in crates or boxes, transported, and handled, usually without refrigeration. The principal postharvest problems are water loss and stem end rots, and both are related directly to physical damage incurred during harvesting. Because of the prevalence and importance of stem end rots, the fruits have been termed "perishable." They can be stored for more than a month at 68°F if the stem tissue has not been damaged at harvest.

In commercial handling, prickly pear fruit are held at temperatures from 50° to 58°F with 86 to 90% relative humidity. Vegetable cladodes can be held for about one week under these conditions.

Sources

Seeds or plants

Abbey Garden, 4620 Carpinteria Ave., Carpinteria, CA 93013 (plants only)

Desert Botanical Garden, 1201 N. Galvin Parkway, Phoenix, AZ 85008 (plants only)

Sacramento Cactus & Succulent Society, Shephard Garden Art Center, 3330 McKinley Blvd., Sacramento, CA 95816

A Sticky Business, Cactus & Succulents, Allan Leroy, P.O. Box 743, Petaluma, CA 94952

More information

Dawson, E. Yale. 1966. *The cacti of California.* University of California Press, Berkeley and Los Angeles, CA.

Everett, Thomas. 1978. *Encyclopedia of horticulture.* The New York Botanical Garden, Rodale Press, Emmaus, PA.

Hunter, Mel. 1985. In defense of *Opuntias. Cactus & Succulent Journal.* vol. 57, September-October, 1985.

Kemp, E. E. 1963. *Cacti and succulents, a practical handbook.* E. P. Dutton and Co., Inc., New York, NY.

Martin, Margaret, P. R. Chapman, and H. A. Auger. 1971. *Cacti and their cultivation.* Charles Scribner's Sons, New York, NY.

Mitich, Larry. n.d. *Prickly pear cactus.* Botany Department Cooperative Extension, University of California, Davis, CA.

Russell, Charles E., and Peter Felker. n.d. *The prickly pears (Opuntia spp.): Plants with economic potential.* Texas A&I University, Kingsville, TX.

Tate, Joyce L. 1978. *Cactus cookbook.* Succulent Cookery International, Cactus and Succulent Society of America.

Prepared by Yvonne Savio and Marita Cantwell de Trejo.

Purslane, Verdolaga

Portulaca oleracea **is a member of the Chenopodiaceae (goosefoot) family.**

There are two cultivated purslane varieties (green leaf and golden) that produce larger leaves than the wild variety. Wild purslane is also edible.

Purslane leaves and stems are very fleshy and succulent. The small, oval, juicy leaves cluster at the end of the smooth, purplish red, prostrate stems that arise from a single taproot. Once sown in a garden, the plants produce flowers containing thousands of tiny black seeds. If left to go to seed, purslane will re-seed itself for repeat crops.

Other names. *Verdolaga* (Spanish); *gwa tsz tsai* (Mandarin Chinese); *ngalog* (Filipino); *pourpier* (French).

Commercially grown purslane near Encinitas in Southern California. (Photo: Hunter Johnson)

Market Information

Use. Whole young plants, and especially young leaves and tender stem tips, can be used as a potherb or eaten raw in salads. The taste is similar to that of watercress or spinach. Seeds can also be eaten raw or ground and made into bread. A 100 gram leaf portion has only 15 calories, but provides more Vitamin A and C and omega-3 fatty acids than most other vegetables.

Culture

Climatic requirements. Purslane grows well in warm weather and is susceptible to frost injury.

Propagation and care. When planted in spring, purslane will flower and fruit in May or June. Purslane grows so rapidly that it can be ready for harvest within 3 weeks of planting. If pulled up and left lying on the ground, purslane will re-root and produce seeds. Heavy irrigation encourages growth.

Close-up of a bunch of purslane tips. (Photo: Hunter Johnson)

Sources

Seed

The Cook's Garden, P.O. Box 65, Londonderry, VT 05148

Le Jardin du Gourmet, P.O. Box 75, St. Johnsbury Center, VT 05863

Thompson and Morgan, P.O. Box 1308, Jackson, NJ 08527

More information

Anon. 1989. *Flower and Garden.* July-August 1989.

Anon. 1988. *Organic Gardening.* vol. 35, no. 6.

Prepared by Keith Mayberry and Claudia Myers.

Quinoa

Chenopodium quinoa is a member of the Chenopodiaceae (goosefoot) family.

Quinoa (pronounced keen´-wah or kwin-o´-uh) belongs to a genus of mostly weedy annuals. Plant heights of different varieties range from 42 to 72 inches. The plant has thick tops. Seed colors range from black to red, orange, yellow, and white. Leaf shapes vary with the variety. The grain must be processed to remove the seed coat, which contains saponin and may be toxic. Quinoa seeds, like millet seeds, are $\frac{1}{16}$ inch in diameter and have two flat surfaces.

Varieties include Cahuil (cuyu), Faro (also Farro; light green foliage, yellow seeds), Linares No. 407 (green with some purple leaves, white-yellow seeds), Milahue, Temuco (yellow-green or a few golden heads, white seeds), D407 (early maturing, semi-dwarf growth habit, yellow compact heads, medium-small kernels), and Isleuga.

A quinoa planting at the University of California. (Photo: Tom Kearney)

Market Information

Current production and yield. Quinoa is a native American crop cultivated for centuries in the high Andes of Peru, Bolivia, and Chile. It was a staple food of the Incas. Primary production occurs in Chile and Peru. In its highest production, quinoa's yield equals that of wheat. The Faro variety performs well in Oregon; the Milahue variety is well-suited to California valleys. The Temuco variety grows well in Washington, California, and New Mexico. The variety Isleuga, a native of Chile, grows successfully in the United States. Several varieties have been grown in the Rocky Mountains, the interior Northwest, and the northern Pacific coast. In Colorado, researchers have obtained yields of 1,200 pounds per acre.

Recent interest in the crop is attributed to its versatility, growth requirements, and nutritional value.

Use. Quinoa can be used as a grain or as a flour. The related species *C. nutalliae* is grown as a vegetable in Mexico. Quinoa's taste has been compared to that of corn, squash, and couscous. It cooks like rice (4 parts liquid to 1 part quinoa) but in half the time, and is used in the same ways. Quinoa expands to 3 to 5 times its original volume, yielding 10 to 12 servings per pound. To toast quinoa, sauté the grain in a frying pan for 10 minutes and then boil in a double quantity of water for 10 to 15 minutes. Quinoa can also be curried, served as a side dish or a meat substitute, and added to salads, soups, breakfast porridges, and puddings. Its flour can be used in baked products, tortillas, and porridge.

Nutrition. Quinoa's composition is 10 to 15 percent protein, 4.5 percent fat, 63 percent carbohydrates, 4.1 percent fiber, 12.6 percent water, and 3.4 percent ash. It is rich in unsaturated oils and is a source of calcium, iron, and essential amino acids.

Culture

Climatic requirements. Quinoa can withstand light frost at 30° to 32°F. When the grain is at the soft dough stage (when the inside of a broken kernel will exhibit a soft starchy or cheesy texture), plants can withstand temperatures as low as 20°F. Temperatures over 95°F cause the vegetative plant to become dormant or lead to pollen sterility.

Propagation and care. Fruiting takes place during periods of short day length. In the Sacramento Valley, the time of sowing is mid-February through April. In Colorado, quinoa has been grown in elevations of 7,000 to 10,000 feet.

Allow distances of 1¼ to 2½ feet between rows for good production. As the distance between plants increases, more panicles are produced to compensate for the smaller number of plants per acre. Plant at a depth of ½ to 1 inch, depending on soil type and moisture. Optimum density is 130,000 plants per acre (¼ to ½ pound of seed per acre).

Although soil preparation does not significantly affect yields, the vigor of the plant decreases with the minimum tillage systems. Seeds germinate within 24 hours and emerge within 3 to 5 days. In Colorado, maturity occurs at 90 to 125 days. By thinning to four plants per foot on the row, you can increase the yield by as much as 25 percent over fields grown without thinning.

Avoid over-irrigation, which causes severe stunting in seedlings. In Colorado, quinoa planted in late April to mid-May may not need irrigation until mid-June if the soil profile was near field capacity at planting. In California, irrigation is required since the crop grows mostly during the dry season.

Harvest and postharvest practices. The yields of plots harvested with self-propelled combine and cutter bar–thresher machinery are lower than those of manually cut fields. Windrowing and combining are not recommended practices to obtain high yields. Hand cutting the plants and laying them on paper to catch the seeds during drying is very effective. Low harvest efficiency and the small amount of seed required for planting will cause a severe problem with volunteers the following season.

Pest and weed problems. Weed control should begin 50 days after seeding. After this point, yields will decrease with each day control is delayed. Pests include flea beetles and many caterpillars. In Colorado, insect pests on quinoa include seedling-feeding, foilar-feeding, stalk- and petiole-tunneling, and seed-feeding insects. The plants are also susceptible to powdery mildew, which causes purple blotching of the leaves.

Sources

Seed

Abundant Life Seed Foundation, P.O. Box 772, Port Townsend, WA 98368

Hudson Seedsman, P.O. Box 1058, Redwood City, CA 94064

Living Tree Centre, P.O. Box 797, Bolinas, CA 94924

Peace Seeds, 2385 SE Thompson Street, Corvallis, OR 97333

Richters, Box 26, Goodwood, Ontario, L0C 1A0, Canada

Seeds Blüm, Idaho City Stage, Boise, ID 83706

More information

Cranshaw, Whiteney S., Boris C. Kondratieff, and Tianrong Qian. Insects associated with quinoa, *Chenopodium quinoa,* in Colorado. *J. Kansas Entomol. Soc.* 63(1):195–99.

Etchevers B., Jorge, and Patricia Avila T. 1981. Efecto de la fecha de siembra, distancia entre surcos y ecotipos sombre di rendimiento y comportamiento de quinoa (*Chenopodium quinoa* Willd) en Chillan (Effect of date of planting, distance between rows and ecotypes on yield and behavior of quinoa [*Chenopodium quinoa* Willd] in Chillan). *Ciencia E Investigacion Agraria.* 8(1):19–26.

Johnson, Duane L., and Robert L. Croissant. 1985. *Quinoa production in Colorado*. Leaflet 112, Colorado State University Cooperative Extension, Fort Collins, CO.

National Academy of Sciences. 1989. *Lost crops of the Incas*. National Academy Press, Washington, DC.

Peréz, Guido Caldern, Mario Blasco Lamenca, and Jess Barboza. 1980. Epocas de siembra y deshierbo para el cultivo dela quinua en condiciones del Altiplano de Puno, Peru. *Turrialba.* 30(2):220–23.

Valiente G., Rafael, Jorge Etchevers B., and Edmundo Hetz H. 1981. Effect of seedbed preparation and seeding and harvesting methods on yield, in Quinoa (*Chenopodium quinoa* Willd). *Agricultura Tecnica.* 41(3):121–26.

Prepared by Tonya Nelson.

Radicchio, Red Chicory

***Cichorium intybus* is a member of the Asteraceae (sunflower) family.**

Radicchio varieties include "forcing" and "nonforcing" types. Forcing types are Red Verona (small, deep red head with flavorful bite at maturity) and Treviso (resembling a small romaine lettuce, with a long, conical red head with white midribs and crisp, tangy leaves). Nonforcing types include Palla Rossa (popular in the United States; dark green exterior leaves, and a head with elongated red leaves and pure white ribs), Castelfranco (semi-heading variety with a loose red and white inner head surrounded by green leaves streaked with rose, pink, green, or bronze in cold weather, milder flavor, more heat resistant); Castelfranco Variegata (crumpled foliage is striped red and yellow), Chioggia (variegated red and white; tighter head than Castelfranco), Giulio (round, compact, red head with very good color; resists bolting), and Cesare.

Radicchio is a red, broadleaf, heading form of chicory. Its leaf colors range from pink to maroon with white midribs; variegations include bronze and almost yellow streaks with green. Some cultivars form loose heads, while others have folded leaves and resemble small cabbages. Leaf texture is similar to but stockier than that of a French endive. The red coloration increases during the colder months.

The first growth of many radicchios is green. The green leaves are tough and very bitter. If these leaves are cut back in fall and the winter is cold, the second growth will be bright red or magenta.

Mature radicchio with the leaves peeled back. (Photo: Charlotte Glenn)

Market Information

Use. Radicchio is a popular European salad vegetable and garnish produced largely in Italy. It has a distinctive, bitter flavor, and is eaten raw or lightly grilled or roasted. Its flavor and color add zest to salads and other dishes. Americans prefer to use the bitter-tasting leaves sparingly. In Italy, there are at least fifteen well-known kinds, from the flat, dark rosettes of Ceriolo to the long, thin leaves of Selvatico da campo to the variegated pink and pale green of Castelfranco. Radicchio also serves as a colorful garnish.

The edible flowers have a faint chicory flavor. They must be used immediately after picking, since they remain open only in the morning hours.

Culture

Propagation and care. Cultural practices are similar to those for endive, escarole, and lettuce, but radicchio requires a longer growing period than lettuce (80 to 85 days in the Salinas Valley) and so may require an extra irrigation. Sow seeds ¼ to ½ inch deep from September through March. In Salinas, radicchio grows on a standard 40-inch bed with a 22- to 24-inch bed top. Two rows grow 12 to 14 inches apart on each bed. Down the row, plants grow every 10 to 15 inches depending on the stand before thinning. Raw seed can be precision seeded with a vacuum-type seeder, but other precision seeders require coated seed. The tops will withstand frost and low temperatures (to 20°F) for short periods. Some radicchio has been transplanted using transplanting modules or plugs.

In the warmer inland valleys, the summer heat can cause bolting and tip burning, so August and September plantings are recommended. Growers should experiment to determine the best planting date for each variety. Some varieties perform best on lighter, better-drained soils. In the Salinas area growers can seed the crop during the summer (March to August) if they use adapted varieties.

Harvest rates are low, sometimes in the 20 to 40% range. Many of the plants will produce either unmarketable heads or no heads at all.

There are two types of radicchio: "forcing" and "nonforcing." Nonforcing radicchio forms a head under normal growing conditions, whereas a forcing variety will form a head only after freezing weather. There are three ways to force radicchio to form a head: (1) cut the leaves off to within 1 inch of the crown 2 to 3 weeks before the first frost, and then dig the roots and store them in a burlap bag in a cool dark place (45° to 55°F) where they will produce a second growth of pale red heads; (2) leave the plants in the ground and cover them with straw or another mulch; or (3) leave the plants in the ground and let the frost kill the outer green leaves. Upon peeling back the dead outer leaves, you will find the red head inside.

Radicchio may be a host for lettuce mosaic. In counties like Monterey that enforce a lettuce-free period for mosaic control, radicchio is also subject to this crop-free period.

Harvest and postharvest practices. The USDA storage recommendation is 32° to 34°F at 95 to 100% relative humidity, with an approximate storage life of 2 to 3 weeks.

Sources

Seed

Abundant Life Seed Foundation, P.O. Box 772, Port Townsend, WA 98368

Bountiful Gardens, 5798 Ridgewood Road, Willits, CA 95490

W. Atlee Burpee & Co., 300 Park Avenue, Warminster, PA 18974

California Gardeners Seed Co., 904 Silver Spur Road, Suite 414, Rolling Hills Estates, CA 90274

Comstock, Ferre & Co., 263 Main Street, Wethersfield, CT 06109

The Cook's Garden, P.O. Box 65, Londonderry, VT 05148

Gleckler's Seedsmen, Metamora, OH 43540

Heirloom Garden Seeds, P.O. Box 138, Guerneville, CA 95446

Johnny's Selected Seeds, 299 Foss Hill Road, Albion, ME 04910

Le Champion Heritage Seeds, P.O. Box 1602, Freedom, CA 95019

Le Jardin du Gourmet, P.O. Box 75, St. Johnsbury Center, VT 05863

Nichols Garden Nursery, 1190 North Pacific Highway, Albany, OR 97321

Park Seed Co., Cokesbury Road, Greenwood, SC 29647–0001

Pinetree Garden Seeds, Route 100, New Gloucester, ME 04260

Redwood City Seed Co., P.O. Box 361, Redwood City, CA 94064

Seeds Blüm, Idaho City Stage, Boise, ID 83706

Shepherd's Garden Seeds, 30 Irene Street, Torrington, CT 06790

Taylor's Herb Gardens, 1525 Lone Oak Road, Vista, CA 92084

Thompson & Morgan, P.O. Box 1308, Jackson, NJ 08527

Vermont Bean Seed Co., Garden Lane, Fair Haven, VT 05743

More information

Chandoha, Walter. 1984. Grow Italian greens—radicchio, escarole, and arugula. *Organic Gardening,* 31(5):80–84.

Eagle Research and Development Inc. 1984. *Radicchio, a salad crop new to the United States.* Eagle Research and Development Inc., Salinas, CA.

Glenn, Charlotte, and Georgeanne Brennan. 1988. *Le Marché Seeds International spring '88 catalog.* L Marché Seeds International, Dixon, CA.

Kline, Roger. 1987. Special vegetables. *Country Journal,* April 1987.

Larkcom, Joy. 1986. Radicchio. *Gardening Magazine.* April 1986.

Mansour, N. S. 1990. *Radicchio.* Vegetable Crops Recommendations. Oregon State University, Corvallis, OR.

Stephens, James. *Minor vegetables.* 1988. Cooperative Extension Bulletin SP-40, University of Florida, Gainesville, FL.

Whealy, Kent. 1988. *Garden seed inventory, 2d ed.* Seed Saver Publications, Decorah, IA.

USDA. 1987. *Tropical products transport handbook.* Agric. Handb. 668. USDA, Washington, DC.

Prepared by Yvonne Savio, John Inman, and Claudia Myers.

Red Currant, White Currant

Ribes rubrum, Ribes sativum, and ***Ribes petraeum*** are members of the Saxifragaceae (saxifrage) family.

Red and white currants are essentially the same fruit, differing only in color. Cultivars come from combinations of three main species of *Ribes,* all of them deciduous shrubs. *Ribes rubrum* is an upright shrub found from northern Europe to Siberia and Manchuria. *Ribes sativum* (*R. vulgare*), which includes the large-fruited Cherry cultivar, is a spreading shrub from the temperate region of western Europe. The vigorous *Ribes petraeum,* which includes the Prince Albert and Goudouin cultivars, is a native of high mountain areas of north Africa and Europe.

Flowers are borne toward the base of 1-year-old stems and on spurs on older stems. Each bud opens into a number of flowers that are joined together on a delicate, drooping stem called a *strig.* Most cultivars have self-fertile flowers, but a few are partially self-sterile.

Ripe red currants, ready to be picked. (Photo: Bernadine Strik)

Market Information

Use. When picked just after they turn red, red currants are unsurpassed for jelly making. They are also used for pies and sauce (sometimes in combination with other fruit) and for wine. The crushed fruit makes a cool, refreshing summertime drink. Some currants can be eaten out of hand if left on the bush for about 3 weeks after the berries first turn red (red cultivars) or translucent (white cultivars). Currants are popular among northern Europeans. They are little known in America because they were banned by federal law in 1920 as a supposed carrier of white pine blister rust. The ban was lifted in 1966, and currants are now enjoying some renewed interest in the United States.

Culture

Climatic requirements. Currants thrive in cool, well-drained fertile soil, in full sun or in partial shade. In warm regions, the bushes prefer heavy soil and should be planted in partial shade or on a north-facing slope. An organic mulch can be used to protect the roots and keep the soil cool and moist.

Propagation and care. Currants are propagated from hardwood cuttings of year-old wood. They usually are grown as bushes spaced 5 feet apart. To grow currants in tree form, remove all but the top three buds

Ripe white currants. (Photo: Bernadine Strik)

from the cutting so sprouts will not grow from below the ground. Set cuttings in the ground in the fall or early spring.

Annual pruning will increase yield and keep plants manageable and healthy. Prune so that most fruits will be borne on spurs of 2- and 3-year-old wood. To maintain a supply of two or three each of 1-, 2-, and 3-year-old stems, use a renewal method of pruning. In the first winter, remove all but two or three stems at ground level. The second winter, remove all but two or three of the stems that grew the previous season. At this point the bush will have two or three each of 1- and 2-year-old stems. Continue this practice every winter. In the fourth winter, cut away any stems more than 3 years old at their bases and shorten long or low-hanging branches.

If you want to grow different cultivars in a small area or against a wall, you can grow currants in cordons as single stems. Plant cordons 1½ feet apart or train them against a wall. To develop a cordon, shorten the single upright stem each winter to 6 inches of new growth and shorten any laterals to two buds. In summer, pinch developing laterals to five leaves as the berries begin to color. When the leader reaches its set height, shorten it each winter to one bud of the previous season's growth, and prune the laterals each winter and summer as before.

Currants have a moderate need for nitrogen and a high potassium requirement. An annual dressing of ½ ounce of actual potassium per square yard will prevent potassium deficiency, which is visible as scorching of the leaf margin. Currants are sensitive to chloride ion toxicity, so muriate of potash (potassium chloride) should not be used.

Pests and diseases. Currants can be grown with little or no spraying. They may require treatments including spraying if such pests as aphids, spider mites, and currant borers cause damage. The imported currantworm, usually a gooseberry pest, can defoliate currant plants quickly. An appropriate insecticide should be applied as soon as currantworm is detected. By cleaning up leaves in autumn, you can help prevent potential disease. Fungicides can be used to control powdery mildew, leaf spot, and anthracnose.

Harvest the whole strig intact unless the fruit is to be used immediately. Ripe currants are very soft and easily injured.

Sources

Plants

NOTE: Red Lake, Wilder, and Minnesota 71 are excellent cultivars and are widely available. Jonkheer van Tets and Cherry are resistant to powdery mildew. The following nurseries offer more extensive selections of cultivars.

Alexander Eppler Ltd., P.O. Box 16513, Seattle, WA 98116-0513

International Ribes Association, c/o Anderson Valley Agricultural Institute, P.O. Box 130, Boonville, CA 95415

Southmeadow Fruit Gardens, Lakeside, MI 49116

Whitman Farms Nursery, 1420 Beaumont NW, Salem, OR 97304

More information

Antonelli, A., et al. 1988. *Small fruit pests—Biology, diagnosis, and management.* Publication EB 1388, Washington State University Agricultural Communications, Pullman, WA.

Baker, Harry. 1986. *The fruit garden displayed.* Cassell Ltd., The Royal Horticultural Society, London.

Darrow, G., and S. Detwiler. 1924. *Currants and gooseberries: Their culture and relation to white-pine blister rust.* Farmer's Bulletin No. 1398. USDA, Washington, DC.

Galletta, G., and D. Himelrick, eds. 1990. *Small fruit crop management.* Prentice Hall Press, West Nyack, NY.

Ourecky, D. K. 1977. *Blackberries, currants, and gooseberries.* Cooperative Extension Publication IB 97. Cornell University, Ithaca, NY.

Reich, Lee. 1991. *Uncommon fruits worthy of attention: A gardener's guide.* Addison-Wesley Publishing Co., Reading, MA.

Thayer, P. 1923. *The red and white currants.* Bulletin 371. Ohio Agriculture Experiment Station.

Prepared by Lee Reich.

Rosemary

Rosmarinus officinalis is a member of the Lamiaceae (mint) family.

Rosemary is a perennial evergreen shrub whose ash-colored, scaly bark and green, needlelike leaves give it an overall gray-green appearance. Pale blue flowers grow in clusters along branches. Fruits are very small, spherical nutlets. Plants can reach 5 to 6 feet outdoors. The fragrance is pungent and piney.

Varieties include the following:

Argenteus: Upright growth, slightly silvery foliage.

Arp: Upright growth, pungent but without a strong and sharp scent, hardy to –10°F.

Aureus: Upright growth, yellow-toned leaves.

Benenden Blue: Upright growth, blue flowers, very narrow leaves.

Blue Boy: Upright growth, blue flowers.

Blue Lagoon: Upright growth, blue flowers.

Blue Sprite: Upright growth, blue flowers.

Collingwood Ingrami: Wood rosemary; upright growth, dark blue flowers, shorter and plumper leaves, looser appearance with gracefully pendulous branches.

Corsicus: Upright growth, blue flowers.

Creeping: Prostrate growth, deep blue flowers, long branches that twist and curl and recurve, blooms almost continuously.

Dutch Mill: Upright growth, slightly silvery foliage.

Erectus: Upright growth, blue flowers.

Fastigate: Upright growth, blue flowers.

Golden Rain: Most compact and vigorous habit of any upright grower, gold and green variegated leaves, dark blue flowers, strong aroma.

Rosemary leaves and flowers. (Photo: Hunter Johnson)

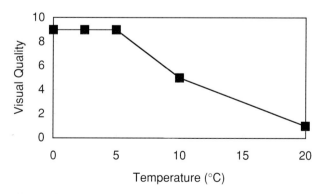

Effect of holding temperature on the quality of fresh rosemary after 1 (●) and 2 (■) weeks' storage in perforated polyethylene bags (both curves are identical, so only one curve is visible here). Visual quality was assessed on a 5-point scale (1, 3, 5, 7, 9, where 1 = low quality and 9 = high quality). From Joyce, Reid, and Katz 1986.

Gorizia: Upright growth, long and broad leaves double the size of other plants, medium blue flowers, gentle aroma, hardy to about 15°F.

Gray: Upright growth, pale blue flowers, grayish foliage, fairly broad leaf.

Humili: Prostrate growth, blue flowers.

Joyce DeBaggio: Upright growth, similar to Golden Rain, hardy to 20°F.

Kenneth: Prostrate growth, blue flowers.

Lockwood de Forest, Lockwoodii, Foresteri, or *Santa Barbara:* Prostrate growth, light blue flowers, lighter green foliage.

Logee Blue: Upright growth, the most intensely blue blooms.

Lottie DeBaggio: Upright growth, strong aroma, sparse foliage, very sensitive to overwatering, hardy to 5°F.

Majorca Pink, Majorca, Roseus, or *Roseus-Cozart:* Upright growth, pink to amethyst violet flowers, produces long branches that twist around the plant and then cascade, blooms sparsely but almost continuously, hardy to 15°F.

Miss Jessup: Upright growth, pale blue flowers, golden green foliage.

Pine Scented: Upright growth, blue flowers, leaves narrow and light green, strong pine scent.

Prostratus or *Lavandulaceus:* Prostrate growth, blue flowers almost continuously, white stems, long branches that twist and curl and recurve.

Rex: Upright growth, blue flowers, very striking appearance, almost black-green leaf color with white beneath, dynamic bloomer.

Roman Vivace: Upright growth.

Seven Seas: Upright growth.

Trailing: Prostrate growth, deep blue flowers, long branches that twist and curl and recurve, blooms almost continuously.

Tuscan Blue: Upright growth, deep blue flowers, reddish stems; leaves light to medium green, glossy, short, and wide, thickly clustered on stems; fast grower, mild fragrance, may reach 10 to 12 feet, hardy to 15°F.

Rosmarinus officinalis *var.* alba *Albus* or *Albiiflorus:* White-flowered rosemary; upright growth, white flowers, dynamic bloomer.

Rosmarinus officinalis *var.* prostratus *Huntington Carpet:* Huntington carpet rosemary; prostrate growth, deep blue flowers, blooms throughout most of the year.

Market Information

Use. Although native to the hills along the Mediterranean, Portugal, and northwestern Spain, rosemary is now widely cultivated. Used fresh and dried, rosemary has medicinal, culinary, dye, and ornamental uses. It is used as an herbal medicine for headaches, muscle spasms, to restore digestion, and for many other conditions. Both the flowers and leaves are used in cooking and as garnishes. Dried or frozen rosemary has a stronger aroma and flavor than fresh. The oil is used to add a pine scent to soaps, lotions, and sachets.

Culture

Climatic requirements. Rosemary can be grown outdoors in climates where winter temperatures do not drop below 27°F (10°F for the *alba* variety). Rosemary is drought tolerant, but good production requires adequate water. Indoors, it is sensitive both to drought and to overwatering.

Propagation and care. Rosemary does well in full sun in well-drained soil with a pH of 6.0 to 7.5. Warm, dry summer climates are ideal. Rosemary does poorly in cold, wet winters unless grown in a protected site. Roots easily develop root rot, so good drainage is essential. Cuttings and layerings from established plants make the best propagation material. Root the cuttings in a mixture of sand, loam, and leaf mold in a cold frame or cool greenhouse. Transplant them 3 feet apart outdoors in the same type of soil. Too much fertilizer will reduce flowering and fragrance, so only fertilize infrequently or if soil is poor.

Rosemary can be started from seed, but germination rates are very low even at optimal conditions of 60°F for 14 to 21 days. It takes up to 3 years to produce a bush large enough for harvesting. Plants grown from seed are not as robust as those propagated from cuttings and layerings.

Harvest. You can harvest throughout the year, but do not remove more than 20 percent of the growth at one time. Volatile oils are most potent just before blooming. To dry leaves, cut individual branches and strip the leaves onto screens or paper in a shady place with good air circulation. You can also dry whole or partial branches for decorative use.

Postharvest handling. Use only the highest-quality plant material for the fresh market. Some of the detailed postharvest handling information provided for basil also applies to rosemary and to other herbs.

The successful marketing of fresh herbs requires careful postharvest handling. Temperature is the most important factor. The optimum postharvest temperature, 32°F, will allow a shelf life of 3 to 4 weeks; 41°F will allow a minimum shelf life of 2 to 3 weeks. Appropriate cooling methods for most herbs include cold rooms, forced-air cooling, and vacuum-cooling. Morning harvest minimizes the need for cooling.

Prevention of excess moisture loss is important. Most herbs respond well to high humidity: relative humidity in the packing area, cold rooms, and transport vehicles should be maintained above 95% where practical. You can also pack the fresh herbs in bags designed to minimize water loss. Maintain constant temperatures and reduce condensation inside the bags to prevent excess moisture and to avoid fungal or bacterial growth. Bags can be ventilated with perforations or fabricated from a polymer that is permeable to water vapor. Young herb tissue is susceptible to ethylene damage. This can be minimized by maintaining recommended temperatures.

If water is used during handling, chlorinated water can reduce the microbial load. To prevent physical injury to leaves, pack them in rigid clear plastic containers or pillow packs.

Sources

Seeds, cuttings, and plants

Abundant Life Seed Foundation, P.O. Box 772, Port Townsend, WA 98368

W. Atlee Burpee & Co., 300 Park Avenue, Warminster, PA 18974

Bountiful Gardens, 18001 Shafer Ranch Road, Willits, CA 95490

Gurney's Seed & Nursery Co., Yankton, SD 57079

Henry Field's Seed & Nursery Co., Shenandoah, IA 51602

Le Jardin du Gourmet, P.O. Box 75, St. Johnsbury Center, VT 05863

Nichols Garden Nursery, 1190 North Pacific Hwy., Albany, OR 97321

Park Seed Co., Cokesbury Road, Greenwood, SC 29647-0001

Shepherd's Garden Seeds, 30 Irene Street, Torrington, CT 06790

Stokes Seeds Inc., Box 548, Buffalo, NY 14240

More information

Cantwell, M., and M. Reid. 1986. Postharvest handling of fresh culinary herbs. *Perishables Handling* No. 60:2–4. Vegetable Crops Dept., University of California, Davis, CA.

DeBaggio, Tom. 1985. Spring is the season to plan your Christmas rosemary crop. *The Business of Herbs.* March-April 1985.

DeBaggio, Tom. 1988. Growing rosemary. *Fine Gardening Magazine.* July-August 1988.

Joyce, Daryl, Michael Reid, and Philip Katz. 1986. Postharvest handling of fresh culinary herbs. *Perishables Handling* No. 58:1–4. Vegetable Crops Dept., University of California, Davis, CA.

Kowalchik, Claire, et al., eds. 1987. *Rodale's illustrated encyclopedia of herbs.* Rodale Press, Emmaus, PA.

Newcomb, Duane, and Karen Newcomb. 1989. *The complete vegetable gardener's sourcebook.* Prentice Hall Press, West Nyack, NY.

Organic Gardening Magazine staff. 1978. *Organic gardening magazine's encyclopedia of organic gardening.* Rodale Press, Emmaus, PA.

Yang, Linda. 1985. Rosemary, right? *Horticulture.* July 1985.

Prepared by Yvonne Savio and Curt Robinson.

Sage

Salvia officinalis is a member of the Lamiaceae (mint) family.

Sage is a hardy perennial shrub that grows from 12 to 30 inches tall. Its woody, wiry stems bear gray-green leaves that are about 2 inches long. The leaves look pebbly and pucker-veined, and may be hairy or velvety. Flowers may be pink, red, blue, purple, pale yellow, or white, and appear from late spring through fall.

Sage varieties include the following:

Salvia officinalis: Broad-leaved, common, or garden sage; used for seasoning.

Salvia o. *Aurea:* Golden sage, 18 inches tall, with striking gold and green variegated leaves, compact and dense growth; a good border plant.

Salvia o. *Dwarf:* Dwarf sage, a very compact grower with smaller leaf size; a good border, rock garden, or container plant.

Salvia o. *Holt's Mammoth:* Holt's mammoth sage, 3 feet tall, larger and rounder leaves than garden sage; grows quickly and is good for cutting and drying in bulk quantities.

Salvia o. *Icterina:* Golden sage.

Salvia o. *Minima:* Dwarf sage.

Salvia o. *Purpurea:* Purple sage, 18 inches tall, compact, aromatic, purple foliage; use like garden sage in stuffings, sausage, omelets, soups, and stews.

Salvia o. *Tricolor:* Variegated, tricolor sage, 2 to 3 feet tall, decorative leaves variegated in cream, purple, green.

Salvia clevelandii: Blue sage, 3 feet tall, blue flowers, used in potpourris and recommended as a substitute for *S. officinalis* in cooking.

Salvia elegans: Pineapple sage, 2 to 3 feet tall, with a pineapple scent, brilliant red flowers; used for drinks, chicken, cheese, jams, and jellies.

Sage growing near Hollister. (Photo: Hunter Johnson)

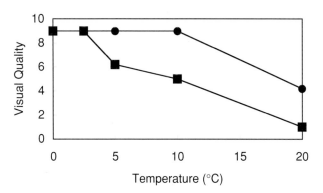

Effect of holding temperature on the quality of fresh sage after 1 (●) and 2 (■) weeks' storage in perforated polyethylene bags. Visual quality was assessed on a 5-point scale (1, 3, 5, 7, 9, where 1 = low quality and 9 = high quality). From Joyce, Reid, and Katz 1986.

Salvia leucantha: Mexican bush sage, 4 feet tall, gray-green foliage, abundant lavender flowers that dry beautifully, not winter hardy.

Salvia sclarea: Clary sage, 3 feet tall, huge gray leaves with spectacular lilac and pink flowers; the most unusual and showy sage.

Market Information

Yield. Yields from second-year growth range from 1,500 to 2,000 pounds of dried sage per acre.

Use. Sage has medicinal, culinary, cosmetic, craft, ornamental, and companion planting uses. It has a lemony, camphor-like, slightly bitter flavor. Sage is a versatile culinary seasoning used in a wide variety of recipes. Sage also contains terpene, camphor, and salvene. Dried sage has a stronger smell than fresh sage. It is an ingredient in perfumes, soaps, and cosmetics. Foliage dries well and can be used in herb wreaths.

Culture

Climatic requirements. Sage does well in full sun in well-drained, moderately rich clay loam soil. It grows best at temperatures of 41° to 79°F, and with a soil pH of 4.2 to 8.3. Sage will withstand temperatures below 0°F if protected by snow or mulch.

Propagation and care. Sage needs moderately high rates of nitrogen. Plants grow best in raised beds where drainage is not a problem. Test the soil for nematodes before planting, since sage is extremely sensitive to them.

Sage seed stores poorly. Conduct a germination test on a small plot before sowing a large amount. Drill seed about ¾ inch deep in rows 3 feet apart as soon as the soil warms in spring. Germination occurs in 7 to 10 days at 60°F. When seedlings are 3 inches tall, thin them to a spacing of 12 to 20 inches. Although a few leaves can be harvested the first year, wait until the next summer for a larger harvest.

Sage is easily propagated from cuttings, divisions, or layerings taken from new growth on established plants in the fall. Transplant them 12 to 20 inches apart in rows 3 feet apart. Water sage well until it is established. Keep the foliage dry and keep the bed well drained. Sidedress plants every spring. Water only sparingly in dry weather.

Harvest. Harvest or prune the plants two or three times from spring through late summer. For full fragrance and flavor, harvest just before the plants bloom. Cut the plants back to about 4 inches above the ground. To dry leaves, snip them from the branches and spread them on cloth or paper out of direct light in a cool, well-ventilated area. They may also be dried in a forced-air drier at temperatures below 125°F. When the leaves are crispy-dry, store them in an airtight, light-tight container. Dried sage has a stronger and slightly different flavor than the fresh. Sage leaves may also be frozen.

To avoid winter killing, make a final, light harvest no later than September. Replace plants after fourth year; otherwise they will become woody and less productive.

Postharvest handling. Use only the highest-quality plant material for the fresh market. Some of the detailed postharvest handling information provided for basil also applies to sage as well as to other herbs.

The successful marketing of fresh herbs requires careful postharvest handling. Temperature is the most important factor. The optimum postharvest temperature, 32°F, will allow a shelf life of 3 to 4 weeks; 41°F will allow a minimum shelf life of 2 to 3 weeks. Appropriate cooling methods for most herbs include cold rooms, forced-air cooling, and vacuum-cooling. Morning harvest minimizes the need for cooling.

Prevention of excess moisture loss is important. Most herbs respond well to high humidity: relative humidity in the packing area, cold rooms, and transport vehicles should be maintained above 95 percent where practical. You can also pack the fresh herbs in bags designed to minimize water loss. Maintain constant temperatures and reduce condensation inside the bags to prevent excess moisture and to avoid fungal or bacterial growth. Bags can be ventilated with perforations or fabricated from a polymer that is permeable to water vapor. Young herb tissue is susceptible to ethylene damage. This can be minimized by maintaining recommended temperatures.

If water is used during handling, chlorinated water can reduce the microbial load. To prevent physical injury to leaves, pack them in rigid clear plastic containers or pillow packs.

Sources

Seeds

Abundant Life Seed Foundation, P.O. Box 772, Port Townsend, WA 98368

W. Atlee Burpee & Co., 300 Park Avenue, Warminster, PA 18974

Bountiful Gardens, 18001 Shaser Ranch Road, Willits, CA 95490

W. Atlee Burpee & Co., 300 Park Avenue, Warminster, PA 18974

The Cook's Garden, P.O. Box 65, Londonderry, VT 05148

Gurney's Seed & Nursery Co., Yankton, SD 57079

Henry Field's Seed & Nursery Co., Shenandoah, IA 51602

Johnny's Selected Seeds, 299 Foss Hill Rd., Albion, ME 04910

Le Jardin du Gourmet, P.O. Box 75, St. Johnsbury Center, VT 05863

Nichols Garden Nursery, 1190 North Pacific Hwy., Albany, OR 97321

Pinetree Garden Seeds, Route 100, New Gloucester, ME 04260

Redwood City Seed Co., P.O. Box 361, Redwood City, CA 94064

Shepherd's Garden Seeds, 30 Irene Street, Torrington, CT 06790

Stokes Seeds Inc., Box 548, Buffalo, NY 14240

Territorial Seed Co., P.O. Box 157, Cottage Grove, OR 97424

More information

Cantwell, M., and M. Reid. 1986. Postharvest handling of fresh culinary herbs. *Perishables Handling* No. 60:2–4. Vegetable Crops Dept., University of California, Davis, CA.

Joyce, Daryl, Michael Reid, and Philip Katz. 1986. Postharvest handling of fresh culinary herbs. *Perishables Handling* No. 58:1–4. Vegetable Crops Dept., University of California, Davis, CA.

Kowalchik, Claire, et al., eds. 1987. *Rodale's illustrated encyclopedia of herbs.* Rodale Press, Emmaus, PA.

Newcomb, Duane, and Karen Newcomb. 1989. *The complete vegetable gardener's sourcebook.* Prentice Hall Press, West Nyack, NY.

Organic Gardening Magazine staff. 1978. *Organic gardening magazine's encyclopedia of organic gardening.* Rodale Press, Emmaus, PA.

Simon, James, Alena Chadwick, and Lyle Craker. 1984. *Herbs: An indexed bibliography 1971–1980.* Archon Books, Hamden, CT.

Whealy, Kent. 1988. *Garden seed inventory, 2d ed.* Seed Saver Publications, Decorah, IA.

Prepared by Yvonne Savio and Curt Robinson.

Salsify, Oyster Plant

Tragopogon porrifolius is a member of the Asteraceae (sunflower) family.
Varieties include Mammoth Sandwich Island and Blanco Améliore.

Salsify is a hardy biennial growing to more than 4 feet in height, with long, smooth, dull green grasslike leaves. It is native to the Mediterranean region. The leaves are flat, 10 to 12 inches long, and taper to a point from a 1-inch base. Roots are white, cylindrical, 8 to 12 inches long, and 1½ to 2 inches in diameter. Purple flower heads as big as 4 inches across form in the spring after planting. Salsify has successfully naturalized as a weed throughout North America.

Spanish oyster plant (also called Golden Thistle), *Scolymus hispanicus,* is a native of southern Europe and requires the same culture as salsify, though its flavor is milder. Leaves are pinnately cut, and flower heads are small and yellow.

Black salsify, *Scorzonera hispanica,* is a perennial popular in central Europe. The roots are black-skinned and may be somewhat slimy, with a creamy-colored interior. The leaves are deeply veined, have wavy margins, and are sometimes used as salad greens. Flowers are yellow and 2 inches in diameter. The plant can be difficult to establish and is seldom grown in the United States.

Market Information

Salsify is a slow seller, but could be profitable for a small farmer if grown in limited quantity. Salsify is available to consumers from late fall through early spring.

Use. Salsify is grown primarily for its edible root, which tastes like oysters. It is usually peeled and can be used like carrots and parsnips; it can be added to soups and stews, parboiled and baked in a casserole with herbs and butter, pureed, battered and fried, or eaten cooked and cold in salad.

Salsify is similar to parsnips in nutritive value, but is a little higher in protein and has only half the calories. One cup of cooked salsify supplies 40 calories, 3.5 grams of protein, 3.8 grams of fiber, 20.4 grams of carbohydrate, 60 mg of calcium, 19 mg of magnesium, 1.7 mg of iron, and 251 mg of potassium. It has very little Vitamin A, B, C, or E. Flower stalks that develop in the second year can be cut and cooked like asparagus if they are harvested before they get woody.

Salsify is grown for its edible roots, which grow 8 to 12 inches long. (Photo: Marita Cantwell de Trejo)

Culture

Climatic requirements. Salsify requires a long growing season of 120 to 150 days to develop marketable-sized roots. Growers in the northern states plant it in the early spring and harvest it in the fall before the ground freezes. In areas with hot summers and mild winters, autumn planting may be best. Salsify is very winter-hardy and can withstand freezing weather without damage to its roots. In fact, freezing has been reported to improve the oyster-like flavor of the root. However, alternate cycles of freezing and thawing should be avoided

Propagation and care. For commercial root production, grow salsify on deep, light, rich soil with a pH between 6 and 8. Fresh manure reportedly causes forking of roots, but you can use composted or well-aged manures. Seed is short-lived, so purchase it fresh each year. Sow seeds ½-inch deep at a rate of 1 ounce per 75 feet of row to ensure good germination. Thin the seedlings to 3 to 4 inches apart in rows 18 to 24 inches apart. Water regularly to maintain uniform soil moisture. Roots may be left in the ground until they are ready to go to market in the fall and winter; they are not sensitive to frost, but may shrivel quickly when stored out of the ground in a low-humidity environment.

Harvest and postharvest practices. The USDA storage recommendation for salsify is 32°F at 95 to 98% relative humidity, with an approximate storage life of 2 to 4 months. Topped salsify is the same as topped carrots in terms of its storage requirements. High humidity is essential to prevent wilting. Perforated-film crate liners can minimize shriveling in storage. *Scorzonera* has a storage life of 6 months at 32° to 34°F and 95 to 98% relative humidity.

Sources

Seed

Salsify seed is widely available. Some sources are listed below:

W. Atlee Burpee & Co., 300 Park Avenue, Warminster, PA 18974

Harris Vegetable and Flower Seeds, Moreton Farm, 3670 Buffalo Road, Rochester, NY 14624

Henry Fields Seed and Nursery Co., Shenandoah, IA, 51602

Johnny's Selected Seeds, Foss Hill Road, Albion, ME 04910

Le Jardin du Gourmet, P.O. Box 75, St. Johnsbury Center, VT 05863

Lockhart Seeds, Inc., P.O. Box 1361, N. Wilson Way, Stockton, CA 95205

Nichols Garden Nursery, 1190 North Pacific Highway, Albany, OR 97321

Seeds Blüm, Idaho City Stage, Boise, ID 83706

Stokes Seeds, Inc., 737 Main St., P.O. Box 548, Buffalo, NY 14240

Taylor's Herb Gardens, 1525 Lone Oak Road, Vista, CA 92084

Thompson and Morgan, P.O. Box 1308, Jackson, NY 08527

Tsang and Ma, P.O. Box 5644, Redwood City, CA 94063

More information

Hardenburg, Robert E., Allen E. Watada, and Chien Yi Wang. 1990. *The commercial storage of fruits, vegetables, and florist and nursery stocks.* Agric. Handb. No. 66. USDA Agricultural Research Service, Washington, DC.

Rubatzky, Vincent, and Mas Yamaguchi. 1997. *World vegetables, 2d ed.* Chapman and Hall, New York, NY.

Stephens, James. *Minor vegetables.* 1988. Cooperative Extension Bulletin SP-40, University of Florida, Gainesville, FL.

USDA. 1987. *Tropical products transport handbook.* Agric. Handb. 668. USDA, Washington, DC.

Prepared by Janet Caprile and Claudia Myers.

Specialty Lettuce

Lactuca sativa is a member of the Asteraceae (sunflower) family.

Specialty lettuce varieties include Butterhead, Leaf, Romaine or Cos, and Batavian types. Butterhead lettuces, also called Bibb (an open, loose head with green outer leaves and yellow inner leaves, such as Boston), include Merveille de Quatre Saisons, Red Perella, Tom Thumb, Buttercrunch, Red Riding Hood, and Esmeralda.

Leaf lettuces have a loose rosette of leaves. This type tolerates heat better than other types and is a summer mainstay in warm weather areas. Varieties include Red Sails, Red Salad Bowl, Waldmans Green, Flame, Green or Red Oakleaf, Lolla Rossa, and Royal Red.

Romaine or Cos lettuces have a tall, slender, upright head. This type withstands heat moderately well, and has more of Vitamins A and C than other types. Varieties include Rouge d'Hiver (good as a baby lettuce), Sucrine, Valmaine, Green Towers, and Paris Island Cos.

Batavian lettuces generally form softer, looser heads than the butterhead types, but retain a crunchy, juicy texture. Varieties include Cybele Batavian, Dorée de Printemps, and Reine des Glaces.

Crisphead or Iceberg lettuce is not a specialty lettuce; it is the familiar supermarket type with a tight, firm ball of leaves. Types include Great Lakes and Vanguard. There are more than 100 varieties.

There are several color distinctions, listed here with example varieties in parentheses: light green (White Boston), medium green (Buttercrunch), dark green (Dark Green Boston), yellow-green (Black-seeded Simpson), full light red (De Morges Braun), full medium red (Red Sails), full dark red (Ruby), red tinge (Brune d'Hiver), an bronzy red tinge (Antina).

Tinge is the red color on the leaf margins of such varieties as Prizehead or Antina. Full red is an overall red coloration, such as with Merveille des Quatre Saisons, even though the inner leaves are a light bright green. All full reds still have green interiors, a quality desired by consumers.

A range of baby lettuce varieties growing in raised beds at Kona Kai Farms, Berkeley. (Photo: Charlotte Glenn)

Green oakleaf lettuce. (Photo: Charlotte Glenn)

Fancy lettuces are now available in every green-red combination in every type of head. Lettuce is native to the eastern Mediterranean area.

Market Information

Specialty lettuce is usually marketed as baby or mid-size: the market for full-size specialty lettuce is more limited. As a "baby," it can be packed in an LA lug or similar-sized box with 48 to 72 heads packed snug. For a nice looking box, pack with four varieties, half of them green and half red. Keep the harvested lettuces wet and cold. Lettuces are often packed in a poly bag inside the box.

Use. In salads, specialty lettuces add color, flavor, shape, and texture.

Nutrition. Specialty lettuces are good sources of fiber and potassium. Romaine and leaf lettuces contain about 40% of the recommended daily allowance (RDA) for Vitamin A in a 3½-ounce serving.

Culture

Climatic requirements. Climate is the major influence on a lettuce crop. Lettuce grows best as a cool-season crop, with temperatures of 70° to 75°F being ideal. Long, warm summer days cause most varieties to bolt (go to seed), leaving the plants tough, bitter, and unusable. Very cold weather may cause blistering, dull green color on the leaves, and head formation in loose-leaf types.

Propagation and care. In central California coastal areas, plant most varieties from seed in February and March or August and September. In Southern California inland valleys and low desert areas, seed from early September through early November. In southern coastal areas, seed from September through February. In the Sacramento Valley, plant in September and October. Lettuce planted in October will hold all winter long through January.

Lettuce should be planted ½ inch apart in rows 10 to 20 inches apart and thinned to 8 inches to 14 inches apart in the row depending on the type. The seed should be planted no deeper than ¼ inch. Germination takes 7 to 10 days. Lettuce can also be grown as transplants and placed in the field 30 to 40 days after emergence. For direct-seeded fields, thin the plants when they have two to three true leaves. Lettuce needs ample supplies of nitrogen and phosphorus and regular irrigations. Lettuce may be harvested at any time. Some cooks make salads from extra-small lettuce plants removed during thinning.

When grown for harvest as baby lettuce, plants are most commonly planted close together in raised beds.

Harvest and postharvest practices. Store at 34°F with 90 to 95% relative humidity. Air circulation is important. Stagger stacking, no more than two layers deep, to attain maximum circulation.

Sources

Seed

Abundant Life Seed Foundation, P.O. Box 772, Port Townsend, WA 98368

W. Atlee Burpee & Co., 300 Park Avenue, Warminster, PA 18974

The Cook's Garden, P.O. Box 65, Londonderry, VT 05148

Le Jardin du Gourmet, P.O. Box 75, St. Johnsbury Center, VT 05863

Johnny's Selected Seeds, Foss Hill Road, Albion, ME 04910

Nichols Garden Nursery, 1190 North Pacific Highway, Albany, OR 97321

Ornamental Edibles, 3622 Weedin Court, San Jose, CA 95132

Park Seed Co., Cokesbury Road, Greenwood, SC 29647-0001

Seeds Blüm, Idaho City Stage, Boise, ID 83706

Shepherd's Garden Seeds, Shipping Office, 30 Irene Street, Torrington, CT. 06790

Stokes Seeds, P.O. Box 548, Buffalo, NY 14240

More information

Glenn, Charlotte, and Georgeanne Brennan. 1988. *Le Marché Seeds International spring '88 catalog.* Le Marché Seeds International, Dixon, CA.

Kline, Roger. 1987. Special vegetables. *Country Journal,* April 1987.

The Packer. 1989. *1989 produce availability and merchandising guide.* Vance Publishing Corp., Overland Park, KS.

USDA. n.d. *Table of container net weights.* USDA Marketing Service, Washington, DC.

Prepared by Nancy Garrison, Wes Foster, Claudia Myers, and Keith Mayberry.

Specialty Mustard

Brassica juncea varieties are members of the Brassicaceae (mustard) family: Chinese Green Mustard, Gai Choy (*Brassica juncea* var. *rugosa* var. *foliosa*); Mizuna, Potherb Mustard (*Brassica juncea* var. *japonica*)

Specialty mustard varieties include Mizuna (usually listed in seed catalogs simply as Potherb Mustard, Mizuna, or Kyona), Chinese Green Mustard (usually listed simply as Gai Choy [*foliosa*], Dai Gai Choy [*rugosa*], Tendergreen, Green-in-Snow, India Mustard, Common Leaf, Leaf Mustard, or AkaTakana); Giant Red Mustard; Osaka Purple; and Fordhook Fancy. The most popular specialty mustards today are the red Asian varieties, such as Giant Red and Osaka Purple.

Several species of *Brassica* are known as mustards; most are annuals grown for the cluster of basal leaves used as greens. According to author Geri Harrington, the Chinese green mustard *rugosa* varieties have brownish red leaves rather than the dark green leaves of the *foliosa* varieties. *Rugosa* also has broader, thicker leaves than *foliosa*.

Mizuna is grown widely in Japan. The plant is 12 to 18 inches tall with yellow-green, smooth, and slightly fuzzy leaves, which have a distinctive shape: deeply notched, narrow, and feathery.

Market Information

Use. The leaves may be eaten raw in a salad mixed with other greens. All of the mustards can be steamed, boiled, or stir-fried. Gai choy varieties can be treated just like spinach.

Culture

Climatic requirements. Specialty mustards should be grown at the same time of year as other cool-season greens. Mizuna withstands frost and light freezes and is not quick to seed, even in periods of warm weather that may occur during the winter months.

Propagation and care. Mustards come up quickly from seeds. Keep them well watered. Mizuna seeds should be sown ½ inch deep in rows 12 to 18 inches apart. Thin to stand 6 to 12 inches apart. The leaves are ready for use any time after 3 weeks. Chinese green mustard *foliosa* varieties have small seeds that should be planted ¼ inch deep ¾ inch apart in rows 12 inches apart; thin them to 4 inch-

Red Asian mustard here is planted thick to be harvested young, when the leaves are 4 to 6 inches tall. (Photo: Charlotte Glenn)

Mizuna mustard grows 12 to 18 inches tall and has distinctive, feathery leaves. (Photo: Charlotte Glenn)

es. They should be harvested when 6 to 8 inches tall. Any taller, and they become too pungent. The *rugosa* varieties should be planted ½ inch deep and 2 inches apart, and later thinned to 10 inches.

If you are going to harvest mustards young (when 4 to 6 inches tall), it is best just to broadcast the seed and rake it in to a depth of less than ½ inch. Such thick plantings must be harvested young.

In warm areas, mustards are very susceptible to flea beetle attacks. Insecticides are not recommended for young salad greens; use row covers instead.

Sources

Seed

American Takii Inc., 301 Natividad Road, Salinas, CA 93906

W. Atlee Burpee & Co., 300 Park Avenue, Warminster, PA 18974

The Cook's Garden, P.O. Box 65, Londonderry, VT 05148

Johnny's Selected Seeds, Foss Hill Road, Albion, ME 04910

Le Jardin du Gourmet, P.O. Box 75, St. Johnsbury Center, VT 05863

Nichols Garden and Nursery, 1190 North Pacific Highway, Albany, OR 97321

Northern Sales Co. Ltd., 5th Floor – 200 Portage Avenue, Winnipeg, Manitoba R3C 3X2, Canada

Park Seed Co., Cokesbury Road, Greenwood, SC 29647–0001

Sakata Seeds, 18695 Serene Drive, Morgan Hill, CA 95037

Seeds Blüm, Idaho City Stage, Boise, ID 83706

Shepherd's Garden Seeds, Shipping Office, 30 Irene Street, Torrington, CT 06790

Sunrise Enterprises, P.O. Box 10058, Elmwood, CT 06110–0058

Tsang and Ma, P.O. Box 5644, Redwood City, CA 94063

NOTE: American Takii and Sakata are Japanese seed companies that are very familiar with mustards.

More information

Harrington, Geri. 1984. *Grow your own Chinese vegetables.* Garden Way Publishing, Pownal, VT.

Liberty Hyde Bailey Hortorium. 1976. *Hortus third.* Cornell University/Macmillan Publishing Co., New York, NY.

Mansour, N. S. 1990. *Mustard.* Vegetable Crops Recommendations. Oregon State University, Corvallis, OR

Stephens, James. *Minor vegetables.* 1988. Cooperative Extension Bulletin SP-40, University of Florida, Gainesville, FL.

Prepared by Claudia Myers.

Specialty Tomatoes

Lycopersicon lycopersicum (*esculentum*) **is a member of the Solanaceae (nightshade) family.**

The tomato plant is native to tropical America, but has been cultivated in Europe and the United States for more than 200 years. There are about 400 varieties available commercially today. Many older varieties have been lost, but breeding programs continue to add new varieties each year. Many of the varieties described here are old varieties that are carried by only a few seed companies.

Market Information

Production and yield. High-volume tomato production and marketing are significantly influenced by price, and competition is fierce. Varieties are selected for yield, tolerance to shipping, fruit size, fruit color, ripening date, and pest resistance.

Exotic tomatoes are both eye-catching and delicious. (Photo: Paul Vossen)

Use. Consumers are most familiar with red tomatoes that are harvested at a mature green stage and allowed to ripen off the vine. There is a market opportunity for flavorful, vine-ripened fruit of many colors as specialty items sold for their uniqueness. Out-of-season tomatoes sold from November to May also command significantly higher prices.

Culture

Climatic requirements. The tomato is a warm-season plant that produces fruit 4 to 5 months after seeding. It does not tolerate frost, and develops leaf diseases under high humidity. Temperatures above 50°F are required for proper fruit set. Optimum performance occurs between 65° to 90°F.

Bragger tomatoes. (Photo: Paul Vossen)

Propagation and care. Specialty tomato varieties are seeded into flats 8 to 10 weeks before the planned outdoor transplant date. Plants can be set out when danger of frost has passed or they can be protected with row covers. Space the plants 2 to 4 feet apart in rows 4 to 6 feet apart. Staking plants to reduce losses from ground-contact disease is cost effective but labor intensive.

Tomatoes are best irrigated with either furrow or drip irrigation. Overhead water can be applied up to first ripening if managed properly. Drip irrigation greatly reduces weed growth.

Brandywine tomatoes. (Photo: Paul Vossen)

Tomatoes have a moderate need for nitrogen fertilizer (100 to 150 pounds per acre per year). Fruit set is delayed by high nitrogen levels, but a deficiency reduces the number of flowers. Too much moisture stress reduces yield and fruit size, but moderate stress can intensify flavor. Large fluctuations in soil moisture contribute to blossom-end rot and fruit cracking.

Harvest and postharvest practices. Fruits should be harvested every 2 to 3 days during the peak of the season when the fruit is still firm and the final color is not quite achieved. Ripe, delicate fruits should be packed in single-layer flats, handled very carefully, and delivered within a few hours to the ultimate consumer.

Pests and diseases. Although many diseases and pests can cause severe damage to this crop, tomatoes are routinely grown on a small scale without pesticides. The growing area should be rotated away from Solanaceous plants (peppers, potatoes, eggplant, tomatillo, and nightshade weeds) for 1 to 2 years. Plant disease-resistant and locally adapted varieties when possible.

Varieties

Performance of a given tomato variety varies from year to year and from one climatic region to another. Planting date, fertility, irrigation, pest control, and staking can greatly affect performance. A particular variety should not be dismissed based on one season's experience. These are some of the more promising specialty tomato varieties:

Big Rainbow: Large, yellow-orange fruit with red streaks in flesh; excellent flavor.

Bingo: Very large red fruit; determinate vine, high yields.

Bragger: Very large, red, mid-season variety with excellent flavor.

Brandywine: Large, deep red-purple fruit; excellent flavor.

Currant: Pea-sized fruit are harvested in clusters. Requires warm climate, and loses flavor if overirrigated.

Early Girl: Medium-sized, red, 5-ounce fruit; prolific, early, and excellent flavor.

Evergreen: Small, 2-inch-plus fruit; low acid, green when ripe.

Garden Peach: Small orange-yellow fruit; fuzzy skin.

Genovese (Costoluto): Very large, uniquely shaped, ribbed scarlet fruit.

Gold Nugget: Cherry-sized, round, golden-yellow fruit, medium flavor.

Golden Boy: Large, mid-season variety; prolific, good flavor, low acid, golden orange color.

Golden Jubilee: Large, mid-season variety; sweet fruit, prolific, golden orange color.

Green Grape: Large cherry-sized, green-yellow fruit with excellent flavor.

Green Zebra: Small 1½- to 2-inch fruit with light green skin turning orange, and overlaid by dark green stripes; very flavorful; susceptible to verticillium wilt; early variety.

Ida Gold: Large cherry-sized tomato with golden orange fruit; very early, prolific.

Lemon Boy: Medium-sized fruit, fair production; indeterminate plant, bright yellow fruit.

Marvel Striped: Very large, orange-yellow solid-colored fruits with red blossom end and red stripes through the flesh; sweet, prolific, early maturing.

Nepal: Medium to large red fruit; excellent flavor.

Pineapple: Large, red-yellow fruit with red streaks.

Pink Girl: Medium to large pink fruit; prolific.

Purple Calabash: Small, novelty, ruffled shape, purple-pink fruit.

Roma: Small, elongated, oval fruit; paste-type flesh, good flavor, very prolific.

San Marzano: Small, red, rectangular paste tomato; excellent flavor.

Sundrop: Cherry-sized, deep orange, meaty fruit.

Sweet 100: Cherry-sized red fruit; grows in clusters, very flavorful, very productive.

Tangerine: Large, flavorful orange fruit.

Taxi: Small to medium-sized, bright yellow, attractive fruit; small but very prolific plant.

Tigerella: Small, early maturing red fruit with orange stripes; prolific.

White Beauty: Small to medium-sized green fruit turns white at maturity; soft, sweet, mild flavor; fruit ripens late.

Whopper: Medium to large red fruit with excellent flavor; good producer.

Yellow Oxheart: Very large, late maturing, heart-shaped fruit; excellent flavor, paste-type flesh.

Yellow Pear: Cherry-sized, pear-shaped, bright yellow fruit; excellent flavor.

Currant tomatoes.

Green Zebra tomatoes.

Pineapple tomatoes.

Gold Nugget cherry tomatoes.

Lemon Boy tomatoes.

Marvel Striped tomato.

Golden Jubilee tomatoes.

Tigerella tomatoes.

Roma tomatoes.

Green Grape cherry tomatoes.

Sundrop cherry tomatoes.

Yellow and red pear cherry tomatoes.

All photographs on this page are by Paul Vossen.

Sources

Seeds

Abundant Life Seed Foundation, P.O. Box 772, Port Townsend, WA 98368

Allen, Sterling, and Lothrop, 191 U.S. Route 1, Falmouth, ME 04105

Archias' Seed Store, P.O. Box 109, Sedalia, MO 65301

Burgess Seed and Plant Co., Dept. 89, 905 Four Seasons Road, Bloomington, IL 61701

W. Atlee Burpee Co., 300 Park Avenue, Warminster, PA 18991-0001

D. V. Burrell Seed Growers, Rocky Ford, CO 81067

Butterbrooke Farm, 78 Barry Road, Oxford, CT 06483

Comstock, Ferre & Co., P.O. Box 125, Wethersfield, CT 06109

The Cook's Garden, Box 65, Londonderry, CT 05148

De Giorgi Co., Inc., 1411 Third Street, Council Bluffs, IA 51502

Earl May Seed & Nursery, L. P., Shenandoah, IA 51603

Early's Farm & Garden Center, Inc., P.O. Box 3024, Saskatoon, Saskatchewan, S7K 3S9 Canada

Farmer Seed and Nursery, Dept. 77 Reservation Center, 2207 East Oakland Ave., Bloominton, IL 61701

Fisher's Garden Store, P.O. Box 236, Belgrade, MT 59714

Glecker's Seedmen, Metamora, OH 43540

Grace's Gardens, 10 Bay Street, Westport, CT 06880

Gurney Seed and Nursery Co., Yankton, SD 57079

Harris Moran Seed Company, 1155 Harkins Rd., Salinas, CA 93901

Henry Field's Seed & Nursery Co., Shenandoah, IA 51602

J. L. Hudson, Seedman, P.O. Box 1058, Redwood City, CA 94064

Ed Hume Seeds, P.O. Box 1450, Kent, WA 98032

Johnny's Selected Seeds, 305 Foss Hill Road, Albion, ME 04910

Jung Seeds & Nursery, 335 S. High Street, Randolph, WI 53957-0001

Kitazawa Seed Co., 1748 Laine Ave., Santa Clara, CA 95051-3012

Lagomarsino Seeds, Inc., 5675-A Power Inn Road, Sacramento, CA 95824

Le Marché Seeds International, P.O. Box 190, Dixon, CA 95620

Letherman Seeds Co., 1221 Tuscarawas Street E., Canton, OH 44707

The Meyer Seed Co., 600 S. Caroline Street, Baltimore, MD 21231

Nichols Garden Nursery, 1190 North Pacific Highway, Albany, OR 97321

Park Seed Co., Cokesbury Road, Greenwood, SC 29647-0001

Peace Seeds, 1130 Tetherow Rd., Williams, OR 97544

Piedmont Plant Co., P.O. Box 424, Albany, GA 31703

Pinetree Garden Seeds, New Gloucester, ME 04260

Porter & Son, Seedsmen, P.O. Box 104, Stephenville, TX 76401-0104

Redwood City Seed Co., P.O. Box 361, Redwood City, CA 94064

Rispens, Martin, & Sons, 3332 Ridge Road (rear), Lansing, IL 60438

Seeds Blüm, Idaho City Stage, Boise, ID 83706

Shepherd's Garden Seeds, 7389 West Zyante Rd., Felton, CA 95018

Southern Exposure Seed Exchange, P.O. Box 158, North Garden, VA 22959

Stark Bros., Louisiana, MO 93353-0010

Stokes Seeds, Inc., P.O. Box 548, Buffalo, NY 14240

Territorial Seed Co., P.O. Box 27, Lorane, OR 97451

Twilley Seeds, P.O. Box 65, Trevose, PA 19047

Vermont Bean Seed Co., Garden Lane, Fair Haven, VT 05743-0250

Wetsel Seed Co., Inc., P.O. Box 791, Harrisonburg, VA 22801-0791

Willhite Seed Co., P.O. Box 23, Poolville, TX 76076

More information

Johnson, Hunter, Jr., Albert W. Marsh, Anthony E. Hall, and Robert F. Kasmire. 1975. *Greenhouse tomato production*. Leaflet 2806. University of California Division of Agricultural Sciences, Berkeley, CA.

Newcomb, Duane, and Karen Newcomb. 1989. *The complete vegetable gardener sourcebook*. Prentice Hall Press, West Nyack, NY.

Rude, Paul, and Larry Strand. 1997. *Integrated pest managment for tomatoes, 4th ed.* Publication 3274. University of California Division of Agriculture and Natural Resources, Oakland, CA.

Whealy, Kent. 1988. *Garden seed inventory, 2d ed*. Seed Saver Publications, Decorah, IA.

Prepared by Paul Vossen.

Sponge Gourd, Chinese Okra, Luffa

Luffa cylindrica (sponge gourd) and *Luffa acutangula* (Chinese okra) are members of the Cucurbitaceae (gourd) family and are thus related to cucumber, muskmelons, and squash.

The Luffa is not well known in the vegetable-growing community and has only minor plantings in California, but the unique nature of the fruits, which are used both for food and industrial purposes, prompts interest in the plants. *Luffa cylindrica,* sponge gourd (also known as smooth gourd, dishrag gourd, loofah, or patola), is grown principally for sponge production, as the name implies. Chinese okra, *Luffa acutangula,* is a food crop that is grown for its immature fruit. Both species reportedly originated in India.

Both species have vigorous climbing vines with yellow flowers. In most cultivars, male and female flowers are produced separately in the axils of the leaves. Some hemaphroditic cultivars (male and female organs in the same flower) are known. Flowers are pollinated by insects, primarily various bee species. The flowering habits of the two species differ: flowers of sponge gourd open during the day whereas the flowers of Chinese okra open in late afternoon and remain open during the night. The flowers of both species are open and receptive to pollination for only one day.

Luffa cylindrica is distinguished from *L. acutangula* by its smooth, cylindrical fruit. *Luffa acutangula* fruit are also cylindrical, but has a tapering neck and 10 prominent longitudinal ridges. Both fruits are generally 2 to 3 inches in diameter and 15 to 18 inches in length when mature.

Other names: *Skoo ah* (Hmong); *hechima* (Japanese); *sinqua* (Cantonese Chinese); *ta tsu kua* (Mandarin Chinese); *patola, cabatiti* (Filipino); and *muop khia* (Vietnamese).

*Luffa gourds on the vine: at top, sponge gourd (*Luffa cylindrica*); above, Chinese okra (*Luffa acutangula*). (Photos: Hunter Johnson)*

Market Information

Use. *Luffa cylindrica* (sponge gourd) is grown mainly for the tough, fibrous netting that remains after the pulpy flesh is removed from mature fruit. The fibrous netting makes excellent sponges that are valued for use in the bath or as dish and pot scrubbers. The spongy material is also used for marine steam engine filters, doormats, table mats, and mattress or shoulder pad stuffing, and as sound-absorbing material.

Mature sponge gourd, ready for harvest. (Photo: Hunter Johnson)

The same kind of spongy network is also present in mature fruit of *Luffa acutangula*, but because the quality is inferior to that of smooth gourd this species is rarely grown for sponge. Immature fruit of this species is known as "Chinese okra" and may be eaten cooked or raw, like summer squash or cucumber. Fruits are harvested 4 to 6 inches long, while they are still tender and before the fibrous network develops.

Yield. Reports from Japan and India indicate that a good yield for sponge production is about 25,000 fruit per acre. With a plant density of 4,400 plants per acre (one plant every 24 inches on rows 60 inches apart), the number of fruit per plant should be limited to five or six.

Culture

Climatic requirements. Luffas are warm-season plants that prefer average monthly temperatures in the range of 65° to 75°F, with daytime highs of 85° to 95°F. The plants are sensitive to frost. Luffas perform well from spring planting anywhere in the southern United States, but reports indicate that successful plantings have been grown as far north as Connecticut and even Maine, so it may be possible to produce Luffa in northern as well as in Southern California.

Propagation and care. Soil and irrigation requirements are similar to those for summer squash, cucumbers, or muskmelons. A deep sandy loam would be ideal, but with proper management successful crops can be grown on any good agricultural soil. Fertilizer requirements of these crops have not been studied, but programs commonly used for other cucurbits should be adequate (e.g., nitrogen at 150 pounds per acre, phosphorus at 50 pounds per acre, and potassium and other elements when soil analysis indicates a need). Apply one-third of the nitrogen and all of the phosphorus before planting, either broadcast and disked in or in a band a few inches to the side and below the plant line. Animal manures can be applied and incorporated a few weeks before planting to supply part of the required soil nutrients. Irrigation practices should be managed to maintain good soil moisture in the top 18 inches of soil where the major root system is located. Drip irrigation can be used successfully on Luffa plantings and provides the most efficient use of water.

Plantings can be established by direct seeding or transplanting. Grow transplants in a way that will not disturb the root system during planting. Plants should be set in the field at the 2- to 3-true-leaf stage. When direct seeding, place the seed 1 inch deep—the objective is to have a final stand of plants spaced about 18 to 24 inches apart in the row. The effect of spacing on fruit size and yield has not been studied, but the suggested spacing is used in countries where these crops are commonly grown. Single rows should be spaced 5 to 6 feet apart. A trellis about 6 feet high is required to support the climbing vine. Stakes should be 3 to 4 feet apart, with a network of vertical stings.

Flowering and fruit set begin approximately 6 weeks after seeding, with warm temperatures and good nutrition. Bee activity in the planting is essential for a good fruit set. When growing for immature fruit (Chinese okra), harvest frequently enough to remove all fruits as they attain marketable size in order to obtain maximum yield. Maturing fruit suppresses the development of younger fruit if left on the plant. Evidence suggests that for sponge production (sponge gourd), it is good practice to limit the number of fruit per plant so as to produce fruit of optimum size. Some of the earliest-setting fruit can be removed, and later-setting fruits may need to be thinned. The relationship between plant spacing, fruit size, and number of fruit per plant is an important consideration in the economics of sponge production. The evidence is not yet clear for luffa, but optimum spacings for other cucurbits have been found to maximize the yield of fruit of the most desirable size.

Pests. All cucurbits are susceptible to many of the same diseases and arthropod pests, such as nematodes, viruses, powdery mildew, leaf miners, and spider mites.

A pesticide used on luffa must be specifically labeled for use on that crop or for cucurbits in general. An amendment to the federal pesticide regulations (June 29, 1983) groups several minor cucurbits including luffa, so that residue tolerances for pesticides labeled for use on cucumbers, muskmelons, and summer squash can be used on luffa if the pesticide manufacturer obtains a group residue tolerance and includes that on the label.

Harvest and postharvest practices. The usual method for preparing the sponge material is to immerse the dry, mature fruit in water for a few days to soften the skin and flesh for easy removal. Other processing methods involve freezing or scalding the fruit. Once cleansed of seeds and flesh, the fibrous network is dried and, for some purposes, bleached in hydrogen peroxide. Different varieties vary in diameter, length, and quality of fiber. Immature fruit of the sponge gourd is rarely used for food in this country, but in India sweet varieties are grown for this purpose.

Like cucumbers and summer squash, immature luffa fruit harvested for food is likely to be susceptible to chilling injury, so refrigeration in storage should not fall below 50°F. Storage conditions have not been studied for luffa fruit, but storage life is probably relatively short. To maintain the best quality, fruit should be marketed within a few days of harvest. Young fruit has tender skin, so care should be taken during harvest and packaging. To prevent damage to the plant, fruit should be cut from the plant rather than pulled.

Sources

Seed

W. Atlee Burpee & Company, 300 Park Avenue, Warminster, PA 18974

Johnny's Selected Seeds, 299 Foss Hill Road, Albion, ME 04910

Sunrise Enterprises, P.O. Box 10058, Elmwood, CT 06110–0058

Tsang and Ma, P.O. Box 5644, Redwood City, CA 94063

More information

Cox, Lilas. 1982. The amazing luffa plant. *Small Far. J.* 6(2):54–5.

Deshpande, A. A., V. M. Bankapur, and K. Venkatasubbaiah. 1980. Floral biology of ridge gourd (*Luffa acutangula* Roxb.). *Mysore J. Agric. Sci.* 14:5–7.

Dillon, Bonnie A. 1979. How to harvest and prepare luffa gourd sponges." *Flower & Garden,* September, 1979.

Ko, B. R., K. B. Lee, J. S. La, S. P. Nho, and K. K. Lee. 1978. Studies on the effect of planting density and level of fertilizer to the growth of *Luffa cylindrica* Roemer." *Hort. Agric. Eng.* 20:47–50.

Martin, Franklin W. 1979. Sponge and bottle gourds, *Luffa* and *Lagenaria*. In *Vegetables for the hot, humid tropics.* Sci. Educ. Admin., USDA, Washington, DC.

Okusanya, O. T. 1978. The effects of light and temperature on the germination and growth of *Luffa aegyptica* (*L. cylindrica*). *Physiol. Plant.* 44(4):429–33.

Okusanya, O. T. 1981. Variations in size, leaf morphology, and fruit characters among 25 populations of *Luffa aegyptica* (*L. cylindrica*). *Can. J. Bot.* 59(12):2618–27.

Okusanya, O. T. 1983. Experimental studies on some observed variations in *Luffa aegyptica* (*L. cylindrica*). *Can. J. Bot.* 61(1):202–10.

Porterfield, W. M., Jr. 1955. Loofah—The sponge gourd. *Econ. Bot.* 9:211–23.

Shah, J. J., Y. J. Thanki, and I. L. Kothari. 1980. Skeletal fibrous net in fruits of *Luffa cylindrica* M. Roem. and *Luffa acutangula* Roxb. Curr. *Trends Bot. Res.* pp. 61–65.

Singh, S. D., and Panjab Singh. 1978. Value of drip irrigation compared with conventional irrigation for vegetable production in a hot arid climate. *Agron. J.* 70(6):945–47.

Sundararajan, S., and C. R. Muthukrishan. 1978. The delicious ribbed gourd." *Indian Hort.* Oct.-Dec.:29.

Tisher, W. G. 1974. Grow your own edible dishcloths. *Org. Gard. & Farm.* 21(6):60.

Prepared by Hunter Johnson, Jr.

Swiss Chard

***Beta vulgaris*, Cicla group is a member of the Chenopodiaceae (goosefoot) family.**

Swiss chard has dark green or red leaves on fleshy, white or red stalks. The large, glossy, crisp leaves can grow as long as 15 inches and as wide as 10 inches. Genetically, Swiss chard is the same family as sugarbeets, fodder beets, and table beets.

Other names. *To jisa* (Japanese); *kwoon taat tsoi, paak tim tsoi* (Cantonese Chinese); *chard* (Filipino); spinach beet; leaf beet; Swiss beet.

Market Information
Baby chard is doing well in the market.

Use. The succulent leaves and midribs are eaten raw or as cooked greens. Sometimes, the fleshy leaf midribs are separated from the leaf blade and prepared much like celery or asparagus. Baby-sized whole plants or individual mature leaves can be used at any stage of maturity.

Culture
Climatic requirements. Swiss chard is a cool-season crop and will withstand severe winters, but also tolerates heat very well. Optimum monthly average temperatures for best growth are 60° to 65°F, with an average minimum of 40°F and an average maximum of 75°F.

Propagation and care. Chard is slow-starting but easy to grow. It may be seeded directly or transplanted from a seedbed or from one point in the row to another. Sow seeds ½ to ¾ inch deep, six to eight plants per foot, in rows 2 to 3 feet apart. The seeds germinate in 7 to 14 days. When seedlings are 4 to 8 inches tall, thin or transplant them to give 6 to 12 inches between plants, depending on whether you plan to harvest baby-sized plants or individual mature leaves.

When rain is scarce or during seasonal dry periods, 14- to 21-day irrigation cycles will ensure continuous yield and quality chard. Chard prefers moist soils, which encourage production of new leaves. Nitrogen fertilizer will also stimulate growth. Like its beet relatives, chard is susceptible to yellowing viruses borne by several aphid species that use crucifers as alternate hosts for aphids and viruses.

Swiss chard has large, glossy, dark green or red leaves with fleshy stalks. (Photo: Marita Cantwell de Trejo)

Harvest and postharvest practices. Once the leaves are 4 to 5 inches long, Swiss chard is mature and ready for harvest. However, the leaves will not achieve full flavor and texture until the chard is about 50 to 60 days old. If the older mature leaves are removed with a sharp knife and the young inner leaves are allowed to remain, the plant will remain productive for as long as a year.

The USDA storage recommendation is 32°F at 95 to 100% relative humidity, with an approximate storage life of 10 to 14 days.

Sources

Seed

NOTE: Swiss chard seed is widely available.

More information

Anonymous. 1984. Leafy greens and celery. *Burpee Gardening Bulletin 2849-8*. W. Atlee Burpee Co. Warminster, PA.

California Agricultural Statistics Service. 1987. *County Agricultural Commissioner data. 1987 annual report.* California Department of Food and Agriculture, Sacramento, CA.

Federal-State Market News Service. 1987, 1988. *San Francisco fresh fruit and vegetable wholesale market prices 1987 and 1988.* California Department of Food and Agriculture Bureau of Market News and USDA Marketing Service.

Organic Gardening Magazine. 1978. *The encyclopedia of organic gardening.* Rodale Press, Emmaus, PA.

Seelig, R. A. 1974. *Fruit and vegetable facts and pointers.* United Fresh Fruit & Vegetable Association, Washington, DC.

USDA. 1987. *Tropical products transport handbook.* Agric. Handb. 668. USDA, Washington, DC.

Prepared by Keith Mayberry, Yvonne Savio, and Claudia Myers.

Tarragon

Artemisia dracunculus is a member of the Asteraceae (sunflower) family.

Tarragon is an aromatic perennial herb native to Europe, southern Russia, and western Asia. The plant reaches a height of 2½ to 4 feet and has thin, erect stems, delicate, narrow green leaves, and greenish white flowers.

Market Information

Use. Tarragon is widely cultivated for its anise-flavored leaves, which are used for seasoning. It is also the source of an aromatic, pungent essential oil called estragon, which can be used as flavoring in pickles and tarragon vinegar. The fresh or dried leaves are used as a culinary seasoning. They are most flavorful and aromatic when used fresh. Tarragon is also used in perfumes, soaps, and cosmetics.

Culture

Climatic requirements. Tarragon grows best in warm, sunny locations on dry soils with good drainage. The plant is intolerant of standing water or poorly drained soils. It grows best at temperatures ranging from 45° to 63°F, with annual precipitation of 1 to 4 feet and a soil pH of 4.9 to 7.8.

Propagation and care. Tarragon is best started from seedlings, divisions, or cuttings. Take divisions in the early spring as new growth appears. Take cuttings in autumn or late spring. Roots spread laterally rather than vertically, so plants must be cultivated carefully, and mulch must be provided over the winter for frost protection. Plants should be divided every 3 or 4 years to reinvigorate growth and flavor. Plants can be multiplied by dividing the crown clumps. Space root divisions 18 inches apart in rows 3 feet apart. Crowns should be subdivided again after 3 or 4 years.

Tarragon plants rarely produce viable seed. Seeds that are available for sale may be the less-versatile false or Russian tarragon (*Artemisia dracunculoides*), or a more vigorous grower that lacks the aromatic oils of French tarragon (*A. dracunculus* var. *sativa*).

Harvesting can begin 6 to 8 weeks after setting out. In summer, harvest the leaves just as the

The tarragon plant reaches 2½ to 4 feet in height and has greenish white flowers and narrow green leaves. (Photo: Charlotte Glenn)

flower buds appear, and continue until late autumn, stopping before the leaves begin to turn yellow. The leaves bruise easily; handle them gently. When the leaves do begin to yellow, cut the plant back to about 3 inches above the ground. Hang harvested stalks in bunches in a warm, dry place out of direct light. The tarragon can be dried, but some color and flavor may be lost.

Postharvest handling. The successful marketing of fresh herbs requires careful postharvest handling. Temperature is the most important factor. The optimum postharvest temperature, 32°F, will allow a shelf life of 3 to 4 weeks; 41°F will allow a minimum shelf life of 2 to 3 weeks. Appropriate cooling methods for most herbs include cold rooms, forced-air cooling, and vacuum-cooling. Morning harvest minimizes the need for cooling.

Prevention of excess moisture loss is important. Tarragon can be held in water. Most herbs respond well to high humidity: relative humidity in the packing area, cold rooms, and transport vehicles should be maintained above 95 percent where practical. You can also pack fresh herbs in bags designed to minimize water loss. Maintain constant temperatures and reduce condensation inside the bags to prevent excess moisture and to avoid fungal or bacterial growth. Bags can be ventilated with perforations or fabricated from a polymer that is permeable to water vapor. Young herb tissue is susceptible to ethylene damage. This can be minimized by maintaining recommended temperatures.

If water is used during handling, chlorinated water can reduce the microbial load. To prevent physical injury to leaves, pack them in rigid clear plastic containers or pillow packs.

Sources

Plants and seeds

French tarragon:

W. Atlee Burpee & Co., 300 Park Avenue, Warminster, PA 18974

Gurney's Seed & Nursery Co., Yankton, SD 57079

Henry Field's Seed & Nursery Co., Shenandoah, IA 51602

Nichols Garden Nursery, 1190 North Pacific Hwy., Albany, OR 97321

Park Seed Co., Cokesbury Road, Greenwood, SC 29647-0001

Mexican tarragon:

Redwood City Seed Co., P.O. Box 361, Redwood City, CA 94064

More information

Cantwell, M., and M. Reid. 1986. Postharvest handling of fresh culinary herbs. *Perishables Handling* No. 60:2–4. Vegetable Crops Dept., University of California, Davis, CA.

Joyce, Daryl, Michael Reid, and Philip Katz. 1986. Postharvest handling of fresh culinary herbs. *Perishables Handling* No. 58:1–4. Vegetable Crops Dept., University of California, Davis, CA.

Kowalchik, Claire, et al., eds. 198. *Rodale's illustrated encyclopedia of herbs.* Rodale Press, Emmaus, PA.

Newcomb, Duane, and Karen Newcomb. 1989. *The complete vegetable gardener's sourcebook.* Prentice Hall Press, West Nyack, NY.

Organic Gardening Magazine Staff. 1978. *Organic gardening magazine's encyclopedia of organic gardening.* Rodale Press, Emmaus, PA.

Simon, James, Alena Chadwick, and Lyle Craker. 1984. *Herbs: An indexed bibliography 1971–1980.* Archon Books, Hamden, CT.

Williams, Louis. 1960. *Drug and condiment plants.* Agric. Handb. 172. USDA, Washington, DC.

Prepared by Yvonne Savio and Claudia Myers.

Thyme

Thymus vulgaris is a member of the Lamiaceae (mint) family.

Thyme is a small, many-branched, aromatic perennial shrub native to the western Mediterranean region. Leaves are lance-shaped and up to ½ inch long. The plant grows up to 1 foot tall and 2 feet wide. Small pink flowers arranged in small clusters appear from spring through fall.

The following are some of the most common of the 150 to 400 *Thymus* species:

T. vulgaris: Broad-leaved, common, English, or French thyme, 6 to 15 inches tall, oval gray-green leaves with tiny white to lilac flowers, small upright shrub.

T. broussonetti: Broussonetii thyme, 5 to 12 inches tall, lavender flowers, shrublet with pine scent.

T. camphoratus: Camphor thyme, 6 to 12 inches tall, compact, dark green leaves, requires mild dry climate.

T. x citriodorus: Lemon thyme, 4 to 12 inches tall, dark green glossy leaves, small bush, used in tea and with fish and chicken.

T. x citriodorus *Argenteus:* silver thyme, 10 inches tall, white-edged leaves, shrublike, good for hanging baskets and as an accent plant.

T. herba-barona: Caraway thyme, 2 to 5 inches tall, shiny dark green leaves, lavender blooms, caraway scent, nice ground cover, good in rock gardens and hanging baskets, and good for flavoring meat, poultry, soup, and vegetables.

T. herba-barona *Nutmeg:* Nutmeg thyme, 4 inches tall, short fat stalks, pink flowers, spicy scent, fast creeper.

T. nummularius: 8 to 12 inches tall, smooth glossy green leaves, bright pink flowers, suitable as a hedge.

T. praecox arcticus, T. serpyllum, *or* T. drucei: Mother-of-thyme, 4 inches tall, dark green leaves, forms a thick and dense mat, traditional ground cover.

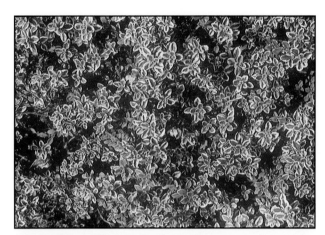

Lemon thyme. (Photo: Charlotte Glenn)

The thyme plant on the left is chlorotic; the plant on the right is healthy. (Photo: Hunter Johnson)

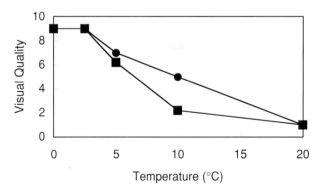

Effect of holding temperature on the quality of thyme after 1 (●) and 2 (■) weeks' storage in perforated polyethylene bags. Visual quality was assessed on a 5-point scale (1, 3, 5, 7, 9, where 1 = low quality and 9 = high quality). From Joyce, Reid, and Katz 1986.

T. praecox arcticus Coccineus: Creeping red-flowered thyme, 4 inches tall, tiny rose-colored blooms, forms dense dark green mat, good as ground cover or between stones.

T. pseudolanuginosus: Woolly thyme, 2 inches tall, minute silver-gray leaves, tiny rose-pink flowers, ground cover among paving stones or along walkways, in rock garden.

T. pulegioides Alba: White moss thyme, similar to Woolly thyme but with abundant white flowers and smoother, bright green leaves.

T. pulegioides Coccineus: Coconut thyme, similar to Woolly thyme, but with tiny dark green leaves and purple-pink blooms.

Others varieties include Doone Valley lemon thyme, dwarf thyme, German winter thyme, golden lemon thyme, golden creeping thyme, and oregano thyme.

Market Information

Use. Thyme has a strong, pungent taste and its leaves and sprigs are used as a culinary seasoning in marinades, vinegars, pickles, and meats. It is also used to flavor liqueurs. Thymol, the essential oil of thyme, has medicinal uses and is used to flavor toothpastes, mouthwashes, and cough medicines. Thymol's antiseptic and stimulating properties are useful in colognes, after-shave lotions, soaps, and detergents.

Thyme flowers, which are edible, may be pink, lavender, crimson, or white. The dried leaves and flowers are used in sachets. The plant is used as a windowsill herb, an outdoor edging plant, and a ground cover.

Yields may go up to 1,500 pounds of dry herb per acre, or 20 pounds of essential oil per acre.

Culture

Climatic requirements. Thyme is extremely adaptable and grows in many regions, from tropical Mediterranean areas to the cold of Scandinavia and Siberia. It will grow with rainfalls from as little as 0.5 inch per year to more than 100 inches, and soil pH ranging from just over 4 to almost 8. Thyme can be grown successfully in any climate having a mean annual temperature anywhere from 45° to 70°F.

Propagation and care. Thyme may be grown from seed or propagated by division. Sow seeds ¼ inch deep and 2 feet apart. Set roots 1½ feet apart in rows 2 feet apart. Growth is slow during early stages. Sidedress with well-rotted compost each spring.

Thyme requires little care and does best in direct sun with little water. Soils rich in nitrogen are most desirable. Well-drained soil is a necessity, since the plant is susceptible to fungal diseases. Protect plants from frost with a mulch. Low, creeping thyme ground covers withstand winters better than bushy varieties, but they are vulnerable to poor soil drainage.

You can take cuttings of new green growth or layer green or woody branches of established plants from mid-spring through early summer. Replace or propagate culinary varieties every 2 or 3 years or they will become woody and produce few tender leaves.

Harvest and postharvest handling. Harvest the leaves of established plants. They are most fragrant on a dry day before noon, just before the plants bloom. Plants can be cut back to 4 inches from the ground. New foliage will fill in again before cold weather, but harvesting a second crop will make the plants less winter hardy. Tie harvested stems in bunches and hang to dry in a warm, airy place, or place stripped leaves on a thin screen to dry, out of direct light. Forced-air dehydrators can be used at 115° to 125°F. Store in an airtight and light-tight container. Leaves can also be frozen.

Use only the highest-quality plant material for the fresh market. Some of the detailed postharvest handling information provided for basil also applies to thyme and to other herbs.

Successful marketing of fresh herbs requires careful postharvest handling. Temperature is the most important factor. The optimum postharvest temperature, 32°F, will allow a shelf life of 3 to 4 weeks; 41°F will allow a minimum shelf life of 2 to 3 weeks. Appropriate cooling methods for most herbs include cold room, forced air cooling, and vacuum cooling. Morning harvest minimizes the need for cooling.

Prevention of excess moisture loss is important. Most herbs respond well to high humidity: relative humidity in the packing area, cold rooms, and transport vehicles should be maintained above 95 percent where practical. You can also pack the fresh herbs in bags designed to minimize water loss. Maintain constant temperatures and reduce condensation inside the bags to prevent excess moisture and to avoid fungal or bacterial growth. Bags can be ventilated with perforations or fabricated from a polymer that is permeable to water vapor. Young herb tissue is susceptible to ethylene damage. This can be minimized by maintaining recommended temperatures.

If water is used during handling, chlorinated water can reduce the microbial load. To prevent physical injury to leaves, pack them in rigid clear plastic containers or pillow packs.

Sources

Seeds

Abundant Life Seed Foundation, P.O. Box 772, Port Townsend, WA 98368

W. Atlee Burpee & Co., 300 Park Avenue, Warminster, PA 18974

Bountiful Gardens, 18001 Shaser Ranch Road, Willits, CA 95490

The Cook's Garden, P.O. Box 65, Londonderry, VT 05148

Gurney's Seed & Nursery Co., Yankton, SD 57079

Indiana Botanic Gardens, P.O. Box 5, Hammond, IN 46325

Nichols Garden Nursery, 1190 North Pacific Hwy., Albany, OR 97321

Park Seed Co., Cokesbury Road, Greenwood, SC 29647-0001

Redwood City Seed Co., P.O. Box 361, Redwood City, CA 94064

Shepherd's Garden Seeds, 30 Irene Street, Torrington, CT 06790

Stokes Seeds Inc., Box 548, Buffalo, NY 14240

Territorial Seed Co., P.O. Box 157, Cottage Grove, OR 97424

More information

Cantwell, M., and M. Reid. 1986. Postharvest handling of fresh culinary herbs. *Perishables Handling* No. 60:2–4. Vegetable Crops Dept., University of California, Davis, CA.

Joyce, Daryl, Michael Reid, and Philip Katz. 1986. Postharvest handling of fresh culinary herbs. *Perishables Handling* No. 58:1–4. Vegetable Crops Dept., University of California, Davis, CA.

Kowalchik, Claire, et al., eds. 1987. *Rodale's illustrated encyclopedia of herbs.* Rodale Press, Emmaus, PA.

Organic Gardening Magazine Staff. 1978. *Organic gardening magazine's encyclopedia of organic gardening.* Rodale Press, Emmaus, PA.

Phillips, Harriet Flannery. 1989. What thyme is it?: A guide to the thyme taxa cultivated in the United States. In *Proceedings of the herbs '89 conference.* International Herb Growers and Marketing Assoc., PA.

Simon, James, Alena Chadwick, and Lyle Craker. 1984. *Herbs: An indexed bibliography 1971–1980.* Archon Books, Hamden, CT.

Talbert, Rex. 1984. Thymus. *The Business of Herbs.* July/August 1984.

Whealy, Kent. 1988. *Garden seed inventory, 2d ed.* Seed Saver Publications, Decorah, IA.

Prepared by Yvonne Savio and Curt Robinson.

Tomatillo

Physalis philadelphica is a member of the Solanaceae (nightshade) family.

The tomatillo is of Mexican origin and is related to the husk tomato. It is an annual, low-growing, sprawling plant usually not more than 2 feet high. The tomatillo has small, sticky, tomato-like fruits enclosed in papery husks. The green or purplish fruits are 1 to 3 inches in diameter.

Varieties. Little breeding work has been done on the tomatillo. The only known cultivar is Rendidora from Mexico, although growers have attempted to improve on it by their own selections. Little has been accomplished, however, and tomatillo remains a highly variable crop in terms of plant habit, fruit size, harvest date, and other characteristics. Tomatillo is primarily a self-incompatible, out-crossed plant, although some self-pollination is known to occur.

Market Information

There are two well-defined markets for tomatillo: organic specialty, and regular bulk commercial. Purple tomatillos are distinctive.

Current production and yield. The tomatillo is an important vegetable crop in Mexico, with over 27,000 acres (11,000 ha) planted, and small plantings are grown in the warmer areas of California. Commercial cropping has been successful along the central and south coasts of California, as well as in the low deserts and the Central Valley.

Use. Tomatillo is widely used as a principal ingredient in green salsa, but is also used in soups and stews. Fruit that begins to yellow is of low culinary value.

Nutrition. The tomatillo's contents of Vitamin A (approximately 30 mg per 100 g) and Vitamin C (less than 4 mg per 100 g) are low compared to those of tomato, a good source of both vitamins. The tomatillo is second only to mushrooms in niacin content.

Culture

Climatic requirements. Like its close relative the tomato, tomatillo is a warm-season crop. It is fairly tolerant of drought.

The tomatillo is a low-growing, sprawling plant. (Photo: Hunter Johnson)

The tomatillo fruit at market maturity. (Photo: Hunter Johnson)

Propagation and care. Tomatillo culture is very similar to that for tomatoes or peppers. Plantings are generally direct-seeded, although tomatillos will transplant well when necessary to fill in stands. Plant spacing and population density vary considerably among growers. Row spacings of 40 inches (single row) to 7 feet (double rows) have been observed. In-row spacings vary from 12 to 36 inches.

Very little work has been done in the United States to refine cultural practices on tomatillos. Cultivation can be tricky because of the plant's low-growing, spreading nature.

Pest and weed problems. Tomatillos can be hit pretty hard by flea beetles. Flea beetle symptoms on tomatillos are similar to those on tomatoes, except that the pests often eat right through the leaves, which are thinner than tomato leaves. Despite heavy damage, infested plants can still produce prolific crops.

Harvest and postharvest practices. First harvest is generally 70 to 80 days from seeding, and the harvest period can exceed 60 days. Fruits are harvested selectively by hand as they attain market size. They should be harvested before they ripen and turn yellow. They are not considered mature until the fruit begins to break through the husk. The husks must be dry at the time of shipping, and that may mean you will need to dry them down after harvest. Preferred fruit size is about 1¼ inches in diameter.

The USDA storage recommendation is 55° to 60°F at 85 to 90% relative humidity, with an approximate storage life of 3 weeks.

Sources

Seed

W. Atlee Burpee & Co., 300 Park Avenue, Warminster, PA 18974

Nichols Garden Nursery, 1190 North Pacific Highway, Albany, OR 97321

Shepherd's Garden Seeds, Shipping Office, 30 Irene Street, Torrington, CT 06790

Native Seeds, 2509 N. Campbell Avenue #325, Tucson, AZ 85719

More information

California Agricultural Statistics Service. 1980–1990. *County Agricultural Commissioner data. 1980 through 1990 annual reports.* California Department of Food and Agriculture, Sacramento, CA.

Cantwell, Marita. 1987. Postharvest handling of specialty crops: Tomatillo. *Perishables Handling.* Vegetable Crops Dept., University of California, Davis, CA.

Dremann, Craig. 1985. *Ground cherries, husk tomatoes, and tomatillos.* Redwood City Seed Company, Redwood City, CA.

Johnson, Hunter. 1985. *Tomatillo.* Vegetable Briefs #244. University of California Cooperative Extension, Davis, CA.

Moriconi, D.N., M.C. Rush, and H. Flores. 1990. Tomatillo: A potential vegetable crop for Louisiana. In Jules Janick and James E. Simon, eds., *Advances in new crops.* Timber Press, Portland, OR.

Morton, J. F. 1987. *Fruits of warm climates.* Creative Resource Systems, Winterville, NC.

Rubatzky, Vincent, and Mas Yamaguchi. 1997. *World vegetables, 2d ed.* Chapman and Hall, New York, NY.

Stephens, James. *Minor vegetables.* 1988. Cooperative Extension Bulletin SP-40, University of Florida, Gainesville, FL.

USDA. 1987. *Tropical products transport handbook.* Agric. Handb. 668. USDA, Washington, DC.

Prepared by Claudia Myers.

Turnip

Brassica rapa, Rapifera group is a member of the Brassicaceae (mustard) family.

Varieties include Purple Top, White Globe, Royal Crown, White Egg, White Lady, Just Right, Tokyo Cross, Milano, and Ohno Scarlet. Seven Top and Topper are grown just for the greens.

Turnips are hardy cool-season plants but are not as frost-resistant as rutabagas. They are biennial, herbaceous dicotyledons grown as an annual for the fleshy roots, which are typically globe-shaped. The flesh color may be white or yellow, but the skin can be white, yellow, red, purple, or black.

Market Information

Marketing. Baby turnips are a specialty crop. There are red turnips and Japanese turnips as well. Turnips should be at least 1 inch in diameter at harvest.

Current production and yield. The approximate yield for a mature root crop of turnips is 5 to 7 tons per acre, or 85 pounds per 100 feet of row.

Use. Both the root and the greens are eaten. They are very popular in Southern states. The baby form (picked before maturity) is a trendy root vegetable. Roots are peeled and then blanched before sautéing, braising, or steaming.

Culture

Propagation and care. Turnip seeds are broadcast or drilled ½ inch deep into rows 10 to 20 inches apart. When seeding to a stand rather than thinning, plant seeds 2 to 4 inches apart in the row, depending on the root size desired. Turnips can also be planted in 4 rows (about 4 inches between rows) on a 40-inch bed, spaced every 2 to 3 inches down the row.

Turnips are a cool-season crop, and can be grown in the same regions and by the same methods used for cole crops. A deep soil with a pH of 5.5 to 7.0 is recommended. Turnip seeds sprout easily in hot weather, so they can be started late in summer. Aphids are the most common pests of turnips. Cabbage maggot is very much attracted to turnips; in fact, turnips have been used with other cruciferous crops as a "trap crop" for cabbage maggots. Flea beetles can cause cosmetic damage by eating small holes in the leafy greens.

Harvest and postharvest practices. The USDA storage recommendation for topped turnips is 32°F at 95% relative humidity, with an approximate storage life of 4 to 5 months. Turnips should be stored in slatted crates with good air circulation around the crates. For marketing, turnips can be placed in perforated plastic bags to help keep humidity high and reduce shriveling.

Turnip greens should be handled like spinach, and stored at 32°F and 95 to 100% relative humidity. Storage life is 10 to 14 days.

Turnip roots have a variety of skin colors, ranging from white to red to black. (Photo: Charlotte Glenn)

Sources

Seed

NOTE: Turnip seed is commonly available.

More information

California Agricultural Statistics Service. 1989. *County Agricultural Commissioner data. 1989 annual report.* California Department of Food and Agriculture, Sacramento, CA.

Federal-State Market News Service. 1988, *San Francisco fresh fruit and vegetable wholesale market prices 1988.* California Department of Food and Agriculture Bureau of Market News and USDA Marketing Service.

Hardenburg, Robert E., Allen E. Watada, and Chien Yi Wang. 1990. *The commercial storage of fruits, vegetables, and florist and nursery stocks.* Agric. Handb. No. 66. USDA Agricultural Research Service, Washington, DC.

Liberty Hyde Bailey Hortorium. 1976. *Hortus third.* Cornell University/Macmillan Publishing Co., New York, NY.

Mansour, N. S. 1990. *Turnip greens.* Vegetable Crops Recommendations. Oregon State University, Corvallis, OR.

Muñoz, Faustino, and Richard Jimenez. 1985. *The gourmet vegetable market in San Diego County.* University of California Cooperative Extension, San Diego County, CA.

Rubatzky, Vincent, and Mas Yamaguchi. 1997. *World Vegetables, 2d ed.* Chapman and Hall, New York, NY.

USDA. 1987. *Tropical products transport handbook.* Agric. Handb. 668. USDA, Washington, DC.

Wechsler, Deborah. 1989. It's turnip time! *National Gardening,* January 1989.

Prepared by Claudia Myers.

Vegetable Amaranth

Amaranthus tricolor (syn. *Agangeticus*), *A. blitum*, *A. dubious,* and *A. cruentus*
are all commonly grown as vegetable amaranth. Most *Amaranthus* species have edible leaves,
especially if they are harvested while the plants are young.

Though characteristics vary among species, the vegetable amaranth plant is generally branched and bushy. Most plants grow to about 18 inches tall, but they can range up to 5 feet. The leaves are oblong and light to dark green and red. All *Amaranthus* species have both male and female flowers on the same plant. Inflorescences consist of a staminate flower clustered with many pistillate flowers. The panicles of vegetable amaranths are much smaller than those of grain amaranths, and most have small black seeds. Like corn, sorghum, and sugarcane, vegetable amaranth benefits greatly from abundant sunlight because it uses light and water so efficiently in converting carbon dioxide to carbohydrate.

Other names. *Hinn choy* (Cantonese); *hsien tsai* (Mandarin); *kalunay, kulitis* (Filipino); *cholai bhaji, lal sag* (Indian); *tampala* (Sri Lanka); and *callaloo* (West Indies).

Market Information

Vegetable amaranth is in demand in early spring for ethnic holiday celebrations, particularly among East Indian and Indonesian consumers. Vegetable amaranth is relatively unknown in the United States, so consumers need to be educated about its use. It is widely cultivated and consumed in Africa, Asia, the Caribbean, and Latin America.

Use. Vegetable amaranth should be prepared as a leafy green vegetable, boiled as you would spinach. Stems and flowers are also edible if harvested at the appropriate time. In Asia and the West Indies, amaranth is used in soup. In Jamaica it is routinely eaten at breakfast and dinner.

Nutrition. The greens are high in calcium, magnesium, iron, Vitamins A and C, and protein.

Culture

Climatic requirements. Amaranth can be grown in tropical or temperate areas. Humid, sunny conditions are advantageous, but not essential. Freezing temperatures will kill the plant.

Leaves of vegetable amaranth are cultivated as edible greens. (Photo: Hunter Johnson)

Propagation and care. Seeds are either sown in narrow rows or broadcast and covered with ¼ to ½ inch of soil. Plants emerge at 7 to 14 days. For a single-harvest crop, the plants are generally thinned to a final spacing of 4 to 6 inches. If plants are to be harvested repeatedly, a wider spacing may be appropriate. Plants removed at thinning can be eaten.

Some growers transplant seedlings, usually 2 to 3 weeks after germination. Transplants should be watered well.

Amaranth requires a fertile soil for abundant growth. Fertilizers, cover crops, or both are recommended for high yields. Good drainage from raised beds can help to alleviate potential root rot problems.

Different varieties of vegetable amaranth have widely varying characteristics. Consider which types are best suited to your soil, climate, and management practices.

Pests and diseases. Grain amaranth varieties in California are susceptible to pythium, and this may also be true of the vegetable varieties. Leafhoppers and Lygus bugs can cause considerable damage.

Harvest and postharvest handling. Practices are similar to those used for other leafy greens such as spinach. Plants can be harvested by cutting or uprooting 4 to 6 weeks after germination. Harvested amaranth should be kept cool and delivered and distributed quickly to market. In Southern California, amaranth can be harvested in as many as four cuttings spaced 2 weeks apart. The exact period to maturity depends upon species, variety, and growing conditions. Most but not all varieties are of best quality when harvested before flowering. If some plants do flower and produce seed, collect and store the dried seed in airtight containers for next season's planting.

Young, tender leaves are sold loose rather than bunched. Pack the leaves as a forest pack.

Sources

Seed

Johnny's Selected Seeds, Foss Hill Road, Albion, ME 04910

Native Seeds/Search, 2509 N. Campbell Avenue #325, Tucson, AZ 85719

Seeds Blüm, Idaho City Stage, Boise, ID 83706

Sunrise Enterprises, P.O. Box 10058, Elmwood, CT 06110-0058

Tsang and Ma, P.O. Box 5644, Redwood City, CA 94063

More information

Daloz, Charles. 1979. Amaranth as a leaf vegetable: Horticultural observations in a temperate climate. In *Proceedings of the second amaranth conference.* Rodale Press, Emmaus, PA.

Feine, Laurie B., Richard R. Harwood, C. S. Kauffman and Joseph P. Senft. 1979. *Amaranth: Gentle giant of the past and future.* Westview Press, Boulder, CO.

Grubben, J. H. 1979. Cultivation methods and growth analysis of vegetable amaranth, with special reference to South-Benin. In *Proceedings of the second amaranth conference.* Rodale Press, Emmaus, PA.

National Research Council. 1984. *Amaranth: Modern prospects for an ancient crop.* National Academy Press, Washington, DC.

Rubatzky, Vincent, and Mas Yamaguchi. 1997. *World vegetables, 2d ed.* Chapman and Hall, New York, NY.

Weber, L. E., et al. 1989. *Amaranth grain production guide.* Rodale Press, Emmaus, PA.

Prepared by Carol Hillhouse.

Water Convolvulus, Chinese Water Spinach, Swamp Cabbage

Ipomoea aquatica is a member of the Convolvulaceae (morningglory) family.
Ipomoea raptans is a synonym for *Ipomoea aquatica*.

Water convolvulus is an herbaceous perennial aquatic or semi-aquatic plant of the tropics and subtropics. Trailing vinelike stems are hollow, adapted for floating in aquatic environments. Roots readily develop at nodes that are in contact with moist soil or water. The succulent foliage and stem tips are light green, and leaves may be arrowhead-shaped, 5 to 6 inches long, or relatively narrow and pointed. Short days favor the development of white and light pink flowers. Purple flowers develop in wild forms of *Ipomoea aquatica*. To obtain seed, allow the flowers to mature and form seed-bearing pods.

There are two major cultivars. Ching Quat has narrow leaves and is usually grown in moist soils. Pak Quat has arrowhead-shaped leaves and is grown in aquatic conditions.

Other names. *Kankon, you-sai* (Japanese); *ong tsoi, weng cai* (Cantonese); *toongsin tsai* (Mandarin); *kang kong* (Filipino, Malaysian); *kang kung, rau muong* (Vietnamese); and *pak bung* (Thai).

Market Information

This fragile crop requires rapid and careful postharvest handling to minimize damage, primarily wilting.

Although water spinach is a highly appreciated vegetable among Asian-American consumers, the scarcity of available aquatic growing areas suited to the crop and the demanding cultural techniques required continue to limit its production in the United States.

Use. Almost all parts of the young plant tissue are edible, but the tender shoot tips and younger leaves are preferred. They are eaten fresh or lightly cooked, like spinach. The leaves and shoot tips are commonly cooked in oil with spices to enhance their bland flavor. Sweet potato shoot tips and leaves resemble and are sometimes substituted for those of water spinach. The coarser stems and leaves from the water spinach plant are often used as animal feed.

Chinese water spinach.
(Photo: Marita Cantwell de Trejo)

Nutrition. Water convolvulus greens provide the nutritional benefits common to most green leafy vegetables. They are high in Vitamins A and C, and have good-quality protein.

Culture

Climatic requirements. The plant is a frost-sensitive perennial with essentially no growth below 50°F. Optimum temperatures for growth are between 75° and 85°F.

Propagation and care. The plants may be grown in moist soils or in aquatic conditions. In moist soils, raised beds are usually used. Seedlings or stem cuttings are transplanted into these beds. Some growers sow seed directly at relatively close spacings (4 to 6 inches) to maximize the yield of tender shoots and leaves. The plants are then grown on trellises to maintain continuing productivity and quality. Shoots regrow readily, and portions of the crop can be harvested several times.

In aquatic plantings, stem cuttings about 12 inches long with 7 to 9 nodes are transplanted into puddled soil, which is prepared as for paddy rice culture. The cuttings are planted about 6 inches deep in rows about 12 inches apart. After transplanting, the field is flooded with continuously flowing water. Water is maintained at a depth of about 2 inches. As growth progresses, the water level is slightly increased. This flooding assists in the control of some weeds. Fertilization is beneficial.

The plants usually grow rapidly in aquatic plantings, with harvest beginning 4 to 5 weeks after transplanting. Semi-aquatic plantings develop more slowly.

Harvest and postharvest practices. Harvested portions should be cleanly cut—not torn—from plants. Once harvested, shoot tips and leaves easily wilt. Cool to 50°F as soon as possible after harvest, and maintain high relative humidities with good ventilation through the packaged product. Avoid bulk packaging. Chilling injury results from holding at temperatures below 40°F for more than 2 or 3 days. To minimize quality loss, the produce should be used soon after harvest.

Pest and disease problems. Production is generally problem-free, although the crop is susceptible to diseases and pests similar to those that affect sweet potatoes. Common insects include whitefly, flea beetles, aphids, and sweet potato weevil. The more troublesome diseases are fusarium wilt or stem rot and *Erwinia* bacterial soft rot.

Sources

Seed

Sunrise Enterprises, P.O. Box 10058, Elmwood, CT 06110-0058

Tsang and Ma, P.O. Box 5644, Redwood City, CA 94063

More information

Rubatzky, Vincent, and Mas Yamaguchi. 1997. *World vegetables, 2d ed.* Chapman and Hall, New York, NY.

Prepared by Vince Rubatzky.

Wax Gourd, Ash Gourd, Winter Melon, Christmas Melon

Benincasa hispida is a member of the Cucurbitaceae (gourd) family.

The pumpkin-like wax gourd vine has thick, furrowed stems with coarse hairs, tendrils, and somewhat triangular, irregularly lobed leaves up to 10 inches long. The flowers are golden yellow and 2½ to 3½ inches wide. The fruits, hairy when young, range from oblong to cylindrical when mature and reach lengths of 16 inches. The mature fruit has a thick layer of white wax (*ton kwa*).

Other names. *Tougan* (Japanese); *doongua, cham kwa* (Chinese); *tankoy* (Filipino).

Market Information

Current production and yield. Most wax gourd is now grown in Southeast Asia, China, and India.

Use. Wax gourd can be cooked as a vegetable when young or made into preserves and sweet pickles when ripe. It is considered a delicacy in Chinese soups.

Culture

Climatic requirements. Wax gourd grows best at 75° to 80°F. Though susceptible to cold, the plant is tolerant of drought.

Propagation and care. Grow the gourds as you would pumpkins or winter squash. Seeds may be planted for spring and fall crops in the desert valleys of Southern California, or as a summer crop in the rest of the state. At UC Davis, May plantings grew very slowly for 2 months, but after 4 months had spread to cover the ground. Very late fruiting and maturation could complicate harvest.

The young fruit of the wax gourd can be cooked and eaten as a fresh vegetable. (Photo: Hunter Johnson)

At maturity, the wax gourd fruit displays a white, waxy bloom. It can grow as long as 16 inches. (Photo: Charlotte Glenn)

Sources

Seed

W. Atlee Burpee & Co., 300 Park Avenue, Warminster, PA 18974

Sunrise Enterprises, P.O. Box 10058, Elmwood, CT 06110–0058

Tsang and Ma, P.O. Box 5644, Redwood City, CA 94063

More information

Liberty Hyde Bailey Hortorium. 1976. *Hortus third.* Cornell University/Macmillan Publishing Co., New York, NY.

Morton, Julia F. n.d. The wax gourd, a year-round Florida vegetable. *Florida State Horticulture Society,* v. 84.

Stephens, James. *Minor vegetables.* 1988. Cooperative Extension Bulletin SP-40, University of Florida, Gainesville, FL.

Prepared by Claudia Myers (adapted from James Stephens's Minor Vegetables*).*

GLOSSARY OF ASIAN VEGETABLES

The authors are Patricia Allen, Administrative Analyst, Agroecology Program, UC Santa Cruz; and Christie Wyman, formerly of the Small Farm Center, UC Davis.

Asian vegetables are popular with California consumers. Because most are grown in small quantities, they offer important opportunities for small farmers.

An Asian vegetable can be known by a variety of names, depending on the language or dialect of the country where the grower, the seller, or even the seed itself originated. This diversity makes it hard to locate and share information about specific vegetables. This glossary will help farmers, farm advisors, researchers, marketers, and consumers identify Asian vegetables.

The glossary is arranged alphabetically by the English name for each vegetable, followed by botanical and family names, Hmong, Japanese, Chinese, Filipino, and Vietnamese names. Many of the Chinese names are followed by an (M) or a (C) to indicate either Mandarin or Cantonese. This list of names is not all-inclusive. We hope to expand and improve it in the future. Please contact the Small Farm Center if you have additions or suggestions.

Thanks to the people who helped with this glossary, including Hunter Johnson, Jr., Vegetable Crops Specialist Emeritus, UC Riverside; Genevieve Ho, Home Economist, Cooperative Extension, Los Angeles County; Mas Yamaguchi, Professor Emeritus, UC Davis; Laura Leonelii, Hmong outreach worker, Sacramento; and Judi Greening, Frieda's Finest Produce Specialties, Inc., Los Angeles.

ENGLISH	BOTANICAL NAME	FAMILY NAME	HMONG	JAPANESE	CHINESE	FILIPINO	VIETNAMESE
Amaranth, edible	*Amaranthus tricolor*	Amaranthaceae		hiyu	hsien shu	kalunay	rao den do
	Amaranthus gangeticus				gien sok hinn choy(C) hsien tsai(M)		
Arrowhead, Chinese	*Sagittaria sagittifolia*	Alismataceae		kuwai	tz'u ku(M) chi goo(C)		
Arrowroot	*Maranta arundinacea*	Marantaceae		kuzi-ukon	chuk shue	aroro	khoai-mi-tinh
Artichoke							
- Chinese/Japanese	*Stachys tuberifera*	Labiatae			kon loh		
- Jerusalem	*Helianthus tuberosus*	Compositae		kiku-imo	chiu sin kaai kuk oo		
Balsam pear: See Melon, bitter							
Bean							
- adzuki	*Phaseolus angularis*	Fabaceae		azuki			
- asparagus	*Vigna unguiculata* var. *sesquipedalis*	Fabaceae	taao-hla-chao	juro-kusasagemae	dow gauk(C)	sitaw	dau-dua
- black gram	*Phaseolus mungo*	Fabaceae		ketsuru-azuki	siu tau		dau den
- broad	*Vicia faba*	Fabaceae		sora-mame	chalm dou(C) chaln dough(M)	patani	
- Chinese long: See Bean, asparagus							
- hyacinth	*Lablab niger*	Fabaceae		fuji-mame	pin tau tseuktau(C)	bataw	dau-vang
- Jack	*Canavalia ensiformis*	Fabaceae		tachinata-mame	dough tau(C)	pataning dagat	
- mung bean	*Phaseolus aureus*	Fabaceae	nong taao pua sha	moyashi-mame	lu tou(M) look dou(C)	balatung	dau-xanh
- mung sprouts	*Phaseolus aureus*	Fabaceae	kolo taac	moyashi	yar tsai(M) ngar choy(C)		
- red	*Vigna umbellata*	Fabaceae			hung tau(C) hoong dough(M)		
- scarlet runner	*Phaseolus coccineus*	Fabaceae			hung fa tsoi tau	parda	
- soy, soya	*Glycine max*	Fabaceae	tau paw	daizu eda mame	dai tou(C) huang dough(M) wong tau(C)	utaw	dau nanh
- yam	*Pachyrrhizus erosus*	Fabaceae	gaw mah sho		di gua(M)		
- yam bean tuber: See Jicama							
- yard-long bean: See Bean, asparagus							
Broccoli, Chinese (white flowering)	*Brassica oleracea* Alboglabra group	Brassicaceae	pak kah nah		gai lohn(C)		
Bur, butter	*Petasites japonicus*	Compositae		fuki	feng tou tsai fung don choi		
Burdock, edible	*Arctium lappa*	Compositae		gobo	niu p'ang(M) ngau p'ong(C)		
Cabbage							
- Chinese	*Brassica rapa* Pekinensis group	Brassicaceae		nappa hakusai	pai-tsai(M) won bok(C)	pechay tsina	
- mustard	*Brassica juncea* var. *rugosa*	Brassicaceae		takana	gai choy(C) jair tsai(M)		cai be xanh
- swamp (water convolvulus)	*Ipomoea aquatica*	Convolvulaceae		kankon	ung choi(C) toongsin tsai(M)	kang kong	kang kung
- white flowering	*Brassica rapa* Chinensis group var. *para chinensis*	Brassicaceae			choi sum(C) paak choi sum(C)		cai be trang
Cassava	*Manihot esculenta*	Euphorbiaceae		tapioka	muk shue	tapioca	khoai-mi
Chayote	*Sechium edule*	Cucurbitaceae	tao tah	hayato uri	fut shau kua(C) tsai hsio li ngow-lai choi(C)	sayote	xu-xu trai su
Chives	*Allium schoenoprasum*	Amaryllidaceae			siu heung tsung		he
Chives, Chinese	*Allium tuberosum*	Amaryllidaceae	ndoh dah	nira	jiu tsai(M) gow choy(C) gil choy		
Chrysanthemum, garland	*Chrysanthemum coronarium*	Compositae		shungiku	tung hao(M) tong ho choi(C)	tunghao	
Coriander, Chinese parsley (cilantro)	*Coriandrum sativum*	Apiaceae	joh tsu	koendoro	yuan sui(C) yim sai(C)	kinchi yun tsai	rao mui
Cucumber, Chinese	*Cucumis sativus*				wong gua(C) huang gua(M)		
Cucuzzi: See Gourd, bottle							
Dandelion, Chinese	*Taraxacum officinale*	Compositae			gow gay(C)		
Dasheen: See Taro							
Daylily	*Hemerocallis fulva*	Liliaceae		kanzou	gum jum(C) huang hua tsai(M)		
Eggplant, oriental	*Solanum melongena*	Solanaceae	tsi lu	nasu	chieh tse(M) ai gwa(C)	talong	ca

ENGLISH	BOTANICAL NAME	FAMILY NAME	HMONG	JAPANESE	CHINESE	FILIPINO	VIETNAMESE
Garlic							
- giant great headed (elephant)	Allium ampeloprasum	Amaryllidaceae		suan	suen(C) suahn(M)	bawang	
- Oriental (Chinese chives)	Allium tuberosum	Amaryllidaceae		nira	su-en(C) shu-an(M)		
Ginger root	Zingiber officinale	Zingiberaceae		shoga	chiang(M) keong(C)	luuya	
Gourd							
- bitter: See Melon, bitter							
- bottle (calabash, white flowered)	Lagenaria siceraria	Cucurbitaceae		yugao	po gua(C)	upo	bau
- fuzzy (harvested at immature stage)	Benincasa hispida	Cucurbitaceae	tao-tue		mao gwa(M) tsit gwa(C)	tankoy	bi-dao
- sponge: See Luffa							
- wax, Chinese preserving melon (white, winter) (harvested at mature stage)	Benincasa hispida	Cucurbitaceae		tougan	doongua(C)(M) cham kwa	tankoy	
Jicama (yam bean)	Pachyrrhizus erosus	Fabaceae	gaw mah sho	kuzu-imo	ti kua(M) sar got(C)	singcamas	cu san
Leek: See Garlic							
Lotus root	Nelumbo nucifera	Nymphaeaceae		hasu renkon	lin ngau(C) ou(M)	baino	sen
Luffa	Luffa acutangula	Cucurbitaceae	skoo ah	hechima	sinqua(C)	patola	muop khia
- angled (Chinese okra, vegetable sponge)					ta tsu kua(M)	cabatiti	
- smooth	Luffa aegyptiaca	Cucurbitaceae			bark gua(C)	bilidan	muop huong
Melon							
- bitter	Momordica charantia	Cucurbitaceae	jee dee ee-ah dee ee-ah	rei shi niga uri	ku kwa(M) fu kwa(C)	ampalaya	muop dang
- Chinese preserving: See Gourd, fuzzy							
- Japanese, oriental pickling	Cucumis melo (var. conomon)	Cucurbitaceae		uri shiro uri	yueh kua yiet kwa		
- winter: See Gourd, wax							
- zucca: See Gourd, bottle							
Mustard							
- broadleaf, Texas	Brassica juncea var. japonica	Brassicaceae	lohng juah	kyona	gai choy(C)	mustasa	
- celery	Brassica rapa Chinensis group	Brassicaceae		chongee	pai tsai(M) pak choi(C)	petsay	
- curled	Brassica juncea var. crispifolia	Brassicaceae		mizuna	tsoi(C)		
- parcel-and-pocket	Brassica juncea var. strumata	Brassicaceae			paau tsing tsoi		
- silver thread	Brassica juncea var. multisecta	Brassicaceae			ngun sz kaai(C)		
- spinach	Brassica rapa	Brassicaceae		unguisu na komatsuma na	bor choi(C) bor tsai(M)		
- swatow	Brassica juncea	Brassicaceae			chiu chau kaai tsoi(C)		
- thousand-veined	Brassica juncea var. multisecta	Brassicaceae			tsing kang tsai(M) ching kong choi(C)		
- wrapped heart, green	Brassica juncea var. rugosa	Brassicaceae			paau sum kaai tsoi(C)		
Nightshade, Malabar	Basella alba	Basellaceae			pu-tin-choi lo kwai		
Okra, Chinese: See Luffa							
Onion, Japanese bunching, Welsh	Allium fistulosum	Amaryllidaceae	ndoh trah	nebuka	chung(C)(M)	sibuyas	hanh-ta
- oriental	Allium chinense	Amaryllidaceae		rakkyo	ch'iao t'ou(M) kil tow(C)		
Parsley							
- Japanese	Cryptotaenia japonica			mitsuba			
- oriental: See Coriander							
Pea							
- pigeon	Cajanus cajan	Fabaceae		ki-mame	muk tau	kadyos	
- snow, sugar	Pisum sativum var. saccharatum	Fabaceae	tau mah	saya-endo	wan tou(M) sic kap woon dou(C) hawlaan tau(C)		
Pepper							
- bell, green	Capsicum annuum	Solanaceae		piiman	ching jil(C) ching jei(M)	sili peaman	ot
- chili, tabasco	Capsicum frutescens	Solanaceae	kua tsao may	tougara-shi	laat tsiu(C)	siling	ot bi
Potato							
- Chinese					see goo		
- sweet	Ipomoea batatas	Convolvulaceae	oh lee ah	satsuma-imo	faan shue(C)	camote	khoai-lang
Pumpkin							
- Japanese	Cucurbita mixta	Cucurbitaceae		kabocha			
- Chinese	Cucurbita pepo				nam kua(C)		

ENGLISH	BOTANICAL NAME	FAMILY NAME	HMONG	JAPANESE	CHINESE	FILIPINO	VIETNAMESE
Purslane	Portulaca oleracea	Chenopodiaceae			gwa tsz tsai(M)	ngalog	
Radish, Chinese, Japanese white, Oriental winter	Raphanus sativus (Longipinnatus group)	Brassicaceae	lo pue	daikon	lor bark(C)	labanos	cu-cai trang
Spinach							
- Ceylon	Basella rubra	Basellaceae	mahn cha leeah				
- Chinese: See Amaranth							
- Indian, Malabar: See Nightshade							
- New Zealand	Tetragonia tetragonioides	Tetragoniaceae			yeung poh tsoi(C)		
- water: See Cabbage-swamp							
Spinach beet (Swiss Chard)	Beta vulgaris Cicla group	Chenopodiaceae		to jisa	kwoon taat tsoi(C) paak tim tsoi(C)	chard	
Squash, Chinese	Cucurbita pepo	Cucurbitaceae	pawg	kabocha	nung gwa	carrabasa	
Sunflower seed	Helianthus annuus	Compositae			heung yat kwai(C) gua tze		
Tampala: See Amaranth							
Tapioca: See Cassava							
Taro	Colocasia esculenta	Araceae-Arum	gah chu- haah	eddo sato-imo koimo	yu tau(M) woo chai(C) woo doi(C)	gabi	khoai-mon
- Violet-stemmed	Xanthosoma violacea	Araceae-Arum					
Tumeric	Curcuma domestica	Zingiberaceae		okon	wong keung(C)	dilaw	
Turnip, oriental	Brassica rapa Rapifera group	Brassicaceae		shogoin	wu tsing(M)		
Water chestnut	Eleocharis dulcis	Cyperaceae		kuro-kuwai	matai(C) pi chi(M)	apulid	cu nang cuau
Watercress	Nasturium officinale	Brassicaceae		kureson	sai yong choi(C) si young tsai(M)		sa-lat xoong
Yam							
- Lisbon, white	Dioscorea alata	Dioscoreaceae		oo-yama imo	pei tsao bak chiu	ubi	khoai-mo
- Japanese/Chinese	Dioscorea batatas	Dioscoreaceae		nago imo	shan yao(M) faan shue(C)		khoai-tu

ANNOTATED BIBLIOGRAPHY

By Ted S. Sibia and Paul Horton, Bio-Ag Department, Shields Library, UC Davis

Introduction

This is a selected bibliography on specialty and exotic crops. No attempt has been made to list all publications available on each crop. The publications listed include only those of a practical nature, covering topics such as propagation, nutrition, climate, soil and water requirements, and cultural aspects. If one or two articles provide complete information on a crop, this is all that will appear listed under the crop name. All of the publications have been reviewed and the majority are available at the UC Davis library. Most county libraries in California will have free access to these publications as well. Any difficult-to-obtain material may be borrowed from the local library through interlibrary loan or by contacting the Small Farm Center at UC Davis, CA 95616, (530) 752-8136.

It is very difficult to make a list of all of the new crops introduced to this country. One of the best ways to identify them and uncover some information is through seed companies; many companies are glad to provide you with any additional information they may have on a crop. Included here is a very select list of seed companies. Another way to keep abreast is to regularly scan the periodicals listed in this bibliography.

Many thanks are extended to Karen Paschke. Her editing and enthusiasm were greatly appreciated. I am also grateful to the Small Farm Center, specifically Claudia Myers, for providing helpful suggestions and encouragement, and also to the LAUC Research Committee, which provided funding for this project.

Ted S. Sibia
Bio-Ag Department
Shields Library
University of California
Davis, CA 95616

BIBLIOGRAPHY

ALLSPICE (*Pimenta dioica; P. officinalis*)
Tall (up to 40 feet) evergreen trees with aromatic foliage; the aromatic fruits are used for the spice.

Devadas, V. S., and T. P. Mohandas. 1988. Studies on the viability of allspice seeds. *Ind. Coc. Areca. Spice. J.* 11(3):99. Presents ways to extend the germination life of a seed that is traditionally planted immediately after harvesting.

Purseglove, J. W., et al. 1981. Pimento. In *Spices*, vol. 1, 290–300. Essex, United Kingdom: Longman. 1981. A detailed discussion of the production methods in Jamaica, many of which are applicable in the United States.

AMARANTH (*Amaranthus* spp.)
National Research Council. 1984. *Amaranth: Modern prospects for an ancient crop*, 14–47. Washington, D.C.: National Academy Press. Basic instructions for cultivating both vegetable and grain amaranth.

Weber, L. E., et al. 1989. *Amaranth grain production guide*. Kutztown, Penn.: Rodale Press.

APIOS (*Apios americana*)
A leguminous tuber that does well on marginal and waterlogged soils. When boiled or fried, it has a flavor between that of potatoes and boiled peanuts.

Duke, J. 1987. Help rediscover an American vegetable: Apios. *Org. Gard.* 34(1):98–101. This article lists tuber sources and a newsletter.

Putman, D. H., G. H. Heichel, and L. A. Field. 1991. Response of *Apios americana* to nitrogen and inoculation. *HortSci.* 26(7):853–55. Added nitrogen and inoculation with soybean rhizobia are both necessary for best yields.

Reynolds, B. D., et al. 1988. Domestication of *Apios americana*. In J. Janick and J. E. Simon (eds.), *Advances in new crops*, 436–42. Portland, Ore.: Timber Press.

APPLES
Vossen, P. 1989. Growing apples organically. *Small Farm News* Jan./Feb.:1, 6–7.

Wood, W. G. 1978. Apples for a mild climate. *CRFG Yearbook* 10:1–2.

APRICOT AND PLUM HYBRID
Staff. 1989. Apricot crossed with a plum?—Apriums, pluots, plumcots. *Sunset* 182(1):140.

ARGULA OR ARUGULA (*Eruca vesicaria*)
Tender, mustard-flavored Mediterranean salad green, fairly fast growing, less bitter if grown in cool fall, winter, and springtime. Used both raw and cooked.

Chandoha, Walter. 1984. Grow Italian greens: Radicchio, escarole and arugula. *Org. Gard.* 31(5):80–84. Although the grower is in the Eastern United States, the basic techniques are presented for organic growing.

ASH GOURD—See *Gourd: Chinese winter melon*

ATRIPLEX
Possible winter and summer forage crop for arid or saline soils; can be irrigated with saline water.

Goodin, J. R. 1979. *New agricultural crops*, 133–48. Boulder, Col.: Westview Press.

Goodin, J. R., and C. M. McKell. 1970. *Atriplex* spp. as a potential forage crop in marginal agricultural areas. *Proc. XIth Int'l. Grass. Congr.*, 158–61.

BABACO (*Carica pentagona*)
A 3- to 6-foot plant with seedless fruit (12 to 15 inches long by 4 to 6 inches wide) with yellow, smooth, thin skin and soft, white, very juicy pulp that is eaten fresh or prepared.

Babaco Section, 1981. *CRFG Yearbook* 13:48–56.

BAEL FRUIT (*Aegle marmelos*)
About the only California limitation is temperatures below 20°F. A 40- to 50-foot tree with variously shaped fruit (2 to 8 inches). The fruit has thin, hard, woody to soft yellowish rind with 8 to 20 triangular segments filled with aromatic, pasty, sweet, resinous pulp. With sugar it can be eaten like a melon or it can be processed into a nectar or marmalade.

Chattopadhyay, P. K., and S. K. Mahanta. 1989. Bael: Media requirements for seed germination and seedling establishment. *Ind. Hort.* 34(3):27. The media used are very common, readily available, and the methods are very simple.

Morton, J. F. 1987. *Fruits of warm climates*, 187–90. Winterville, N.C.: J. F. Morton.

Roy, S. K. 1990. Bael. In T. K. Bose and S. K. Mitra (eds.), *Fruits: Tropical and subtropical*, 740–46. Calcutta: Naya Prokash. A detailed publication from India, which is appropriate since bael is a native fruit of that country.

BAMBOO SHOOTS

Farrelly, D. 1984. *The book of bamboo*, chaps. 8 and 9. San Francisco: Sierra Club.

Kaule, K., and A. Kaule. 1976. Easy-to-grow edible bamboo. *Org. Gard.* 23(7):72–73.

Rubatsky, Vincent, and Mas Yamaguchi. 1997. *World vegetables: Principles, production, and nutritive values* (2nd ed.). New York.: Chapman and Hall Publishing Co.

BARBADOS CHERRY (*Malphighia puncifolia*)

Temperatures below 28°F kill the tree. Has the highest Vitamin C content of any known fruit. Large, bushy shrub up to 20 feet tall and wide tree with three-lobed ½- to 1-inch-wide bright-red, thin-skinned fruit with orange, juicy, acid to subacid flesh. Pulp can be eaten as is, stewed with sugar as a dessert, or be made into syrup or a sauce for cake topping and ice cream.

Morton, J. F. 1987. *Fruits of warm climates*, 204–7. Winterville, N.C.: J. F. Morton.

BEANS

Adzuki (*Phaseolus angularis*)

Harrington, G. 1978. *Grow your own Chinese vegetables*, 73–75. Pownal, Vt.: Garden Way Publishing. A dry bean with a different, slightly sweet flavor, used in desserts and for sprouts.

White, D. 1972. Adzuki bean potential being assessed. *N. Z. J. Agric.* 125(3):43, 45. Note that the seasons are reversed in New Zealand.

Blackeyed peas (*Vigna sinensis*)

Sallee, W. R., and F. L. Smith. 1969. *Commercial blackeye bean production in California*. Calif. Agri. Exper. Stat., Circular 549. Designed for large-scale dry beans but easily adapted to fresh, green.

Fava, broad (*Vicia faba*)

A vegetable grown over the winter for early spring beans.

Hickman, G., and M. Canevari. 1989. Fava beans. *Small Farm News* (May/June):1, 3.

Green and dry

Whealy, Kent. 1988. *The garden seed inventory*. Decorah, Ia.: Seed Saver Publications. For unusual nonhybrid varieties available in 1987. For hybrids and post-1987 nonhybrid varieties, browse general and specialty seed catalogs.

Guar (*Cyamopsis tetragondloba*)

A very high-protein, Indian subcontinent legume of variable height and branching habits. The immature green pods (harvesting starts at 50 to 80 days and continues for a fairly long time) are used like French beans or fried like potato chips. The whole plant is used for cattle forage, green manure, or cover crop, and the gum extracted from ripe seeds is used in foods and other areas.

Whistler, R. F., and T. Hymowitz. 1979. *Guar: Agronomy, production, industrial use and nutrition*, 57–95. W. Lafayette, Ind.: Purdue University Press.

Lablab or hyacinth (*Dolichos lablab*)

An Indian native used throughout the subcontinent as both a dried and fresh pod bean, it is either bushy or twining; the latter grows rapidly to 10 feet on trellises, walls, etc.

Chakravarty, A. K. 1986. Dolichos bean. In T. K. Bose and M. G. Som (eds.), *Vegetable crops in India*, 524–28. Calcutta: Naya Prokash.

Shivashankar, G., and R. S. Kulkarni. 1989. Field beans (*Dolichos lablab*). *Ind. Hort.* 33(4) and 34(1):24–25, 27. More cultivation practices.

Vijay, O. P. and A. Vani. 1990. Early bushy Dolichos bean gives early high yield at low cost. *Ind. Hort.* 35(2):14–17. Could be adaptable here as a green bean.

Mung (*Vigna radiata*)
Harrington, G. 1978. *Grow your own Chinese vegetables*, 79–81. Pownal, Vt.: Garden Way Publishing.

Winged (*Psophocarpus tetragonolobus*)
A 6- to 8-foot climbing vine producing uniquely flavored, high-protein, four-sided pods from late August through September and marble-sized tubers in October. Both winged and yard-long beans used in stir-fry, stews, and salads.

Harrington, G. 1978. Asparagus pea. *Grow your own Chinese vegetables*, 110–113. Pownal, Vt.: Garden Way Publishing.

Yard-long (*Vigna sinensis* var. *sesquipedalis*)
Harrington, G. 1978. *Grow your own Chinese vegetables*, 86–88. Pownal, Vt.: Garden Way Publishing.

Knott, J. E., and J. A. Deanon. 1967. *Vegetable production in Southeast Asia*, 67, 70, 72–74, 76, 79, 82, 86. Los Baños: University of the Philippines College of Agriculture.

BERGAMOT (*Citrus bergamia*)
Traditionally the peel oil is used for perfume. The fruit juice can be used to acidify orange juice or make lemon juice less acidic. Although its exact climatic adaption is still unknown, UC Riverside budwood is available from three nurseries in Tulare, Kern, and San Diego counties.

Maranto, J. 1990. Bergamot: A citrus with a special fragrance. *Calif. Grower* 14 (2):28, 30.

Hodgson, R. W. 1967. Horticultural varieties. In W. Reuther, et al. (eds.), *The citrus industry*, vol.1, 494–96. Berkeley: University of California Division of Agricultural Sciences.

BERRY

Blackberries (*Rubus* spp.) and boysenberries (*Rubus ursinus* var. *loganobaccus*)
Costello, L. R., et al. 1983. *Fruit, nut and grape varieties for home orchards: San Mateo and San Francisco counties*. University of California Cooperative Extension, Leaflet 21338.

La Vina, Paul D. 1982. *Growing boysenberries and olallie blackberries*. University of California Cooperative Extension, Leaflet 2441. A clear presentation of the basics.

Blueberry (*Vaccinium* spp.)
Chaplin, L. T. 1992. Sing a song of blueberries. *Org. Gard.* 39(3):83–86, 88–89. Detailed presentation on organic growing, including soil preparation and amendments, varieties, pruning, diseases.

Costello, L. R., et al. 1983. *Fruit, nut and grape varieties for home orchards: San Mateo and San Francisco counties*. University of California Cooperative Extension, Leaflet 21338.

Darrow, G. M., and J. N. Moore. 1966. *Blueberry growing*. USDA Farmers' Bulletin 1951.

Eck, P. 1988. *Blueberry science*, chaps. 3, 6, 7, 8, 11. New Brunswick, N.J.: Rutgers University Press. Contains new information, and the new cultivars in chapter 3 should help expand the growing areas. Use the information with the University of California Cooperative Extension Leaflet to figure out chilling and other requirements.

Eck, P., and N. F. Childers. 1966. *Blueberry culture*, chaps. 5, 7, 8, 11. New Brunswick, N.J.: Rutgers University Press. Covers all the basics in detail.

Currant and gooseberries (*Ribes* spp.)
Both grow 3 to 5 feet high and wide, many stalked bushes that, depending on the variety, may be thorny and produce ball-shaped fruit ranging from blueberry to cherry-tomato size. Fruit may be very sour to slightly sweet and is eaten raw with sugar and used in soups, sauces, jams, and preserves.

Childers, N. F. 1983. Currant and gooseberry culture. *Modern Fruit Science*, 507–11. Gainesville, Fla.: Horticultural Publications.

General
Barclay, Leslie, Carlton Koehler, and J. B. Bailey. 1982. *Insect and disease management for home berry plantings*. University of California Cooperative Extension, Leaflet 21320.

Huckleberry, evergreen (*Vaccinum ovatum*)
Naturally occurring in coastal forested areas, the berries are used in pies, jams, syrups, and jellies. The foliage is used in floral arrangements.

Miller, R. A. 1987. The culture and harvest of coast huckleberry foliage. *Herb Mark. Rep.* 3(3):1–4.

Lingonberry (*Vaccinium vitis-idaea*)
The somewhat sour red berry of these acid-loving plants is used for preserves and syrups, and at least one of the new varieties is eaten out of hand.

Pliska, K., and K. Scibisz. 1989. Effect of mulching and nitrogen fertilizer upon growth and yield of lingonberries. *Acta Hort.* 241:139–44.

Butkus, V., et al. 1989. Effect of mulching on growth and fruiting of cultivated lingonberries and bogblueberries. *Acta Hort.* 241:265–69. Research is just beginning, so the results differ somewhat between Poland and Lithuania, but the publication emphasizes some basic cultural practices.

Labokas, J., and D. Budriuniene. 1989. Vegetative propagation of lingonberry. *Acta Hort.* 270–72.

Zillmer, A. 1985. Account of three types of *Vaccinium vitis-idaea*. *Acta Hort.* 165:295–97.

Raspberries (*Rubus idaeus*)
Raspberry pruning for optimal production and plant health is complex because of variations among cultivars. Interspecific hybrid cultivars like the 'Royalty' need their pruning requirements studied individually.

Costello, L. R., et al. 1983. *Fruit, nut and grape varieties for home orchards: San Mateo and San Francisco counties.* University of California Cooperative Extension, Leaflet 21338.

Gundersheim, N. A., and M. A. Pritts. 1991. Pruning practices affect yield, yield components and their distribution in 'Royalty' purple raspberry. *J. Am. Soc. Hort. Sci.* 116(3):390–95.

Lawrence, F. J. 1979. *Growing raspberries.* USDA Farmers' Bulletin 2165.

Salalberry (*Gaultheria shallon*)
Native coastal berry from Santa Barbara to British Columbia.

Miller, R. A. 1989. The culture and harvest of salal. *Herb Mark. Rep.* 5(10):4–7.

Schenk, G. 1990. Salalberry: The once and future fruit. *IC&RF Soc. Newsl.* 32 (Winter):8–12.

Saskatoon (*Amelancher alnifolia*)
Tall (up to 18 feet) shrub or small tree producing clusters of mostly blue-purple, blueberry-flavored fruits (up to ⅝-inch in diameter) that are eaten fresh, canned, frozen, and in pies.

Harris, R. D. 1972. *The saskatoon.* Canada Department of Agriculture, Publication 1246. Like the University of California publications, it covers the field and includes the names and brief descriptions of cultivars.

Serviceberry (*Amelanchier* spp.)
Cubberly, B., and E. R. Hasselkus. Amelanchiers: Trees and shrubs with year-round enchantment. *Am. Nursery* 165(9):111–12, 116–17. Article describes varieties and cultivars, including edibility.

Dirr, M. A. 1987. Native amelanchiers: A sampler of northeastern species. *Am. Nursery* 166(5):73–82.

McTavish, B. 1986. Seed propagation of some native plants is surprisingly successful. *Am. Nursery* 164(4):60, 62.

BITTER MELON—See *Gourd: Bitter melon*

BITTERLEAF (*Vernonia calvoana*)
Being tested as an annual in Missouri for specialty ethnic market.

Fube, Dr. H. N., and B. Djonga. 1985. Tropical vegetables in human nutrition: A case study of Ndole. *Acta Hort.* 198:199–205.

BLACKEYED PEAS—See *Beans: Blackeyed peas*

BROCCOLI, CHINESE—See *Cabbage, Chinese*

BUFFALO GOURD
Note: Needs processor and starch. Good potential in arid and semiarid regions, this perennial produces vegetable oil, protein, and starch.

Bemis, W. P., et al. 1978. The buffalo gourd: A new potential horticultural crop. *HortSci.* 13:235–40.

Bemis, W. P., et al. 1979. The buffalo gourd. In Gary A. Ritchie (ed.), *New agricultural crops,* 65, 73–87. Boulder, Col.: Westview Press.

BURDOCK, GOBO (*Arctium lappa, A. minus*)
The stems, new sprouts, and primarily the roots are used as vegetables in Japan and China. The root can also be processed for pharmaceutical and cosmetic uses. Most farmers consider it a weed.

Miller, R. A. 1987. Burdock: A new product for America. *Herb Mark. Rep.* 3(1):1–2.

Halpin, A. 1987. *Domestic burdock in unusual vegetables: Something new for this year's garden,* 69–71. Emmaus, Penn.: Rodale Press.

CABBAGE, CHINESE (*Brassica pekinensis, B. chinensis, B. perviridis*)
Harrington, G. 1978. The Chinese: Celery cabbage, mustard cabbage, flowering cabbage, Chinese broccoli. *Grow Your Own Chinese Vegetables,* 94–109. Pownal, Vt.: Garden Way Publishing.

CACTI
Hegyl, H. 1971. The edible fruited cacti. *CRFG Yearbook* 3:39–50.

Nopales, Cactus pad
Russell, C. E., and P. Felker. 1987. The prickly pears: A source of human and animal food in semiarid regions. *Econ. Bot.* 41:433–45.

Savio, Y. 1987. Prickly pear cactus: The pads are "nopales," and the fruits are "tunas"; they are easy to grow and wonderful to eat. *Cact. Succulent J.* 59(3): 113–17.

CALAMONDIN (*Citrus madorensis*)
Good for marmalade or replacing lemon juice, not for eating out of hand.

Morton, J. F. 1987. *Fruits of warm climates,* 176-78. Winterville, N.C.: J. F. Morton.

CAPE GOOSEBERRY—See *Poha berry*

CAPERS—See *Herbs: Capers*

CARAMBOLA (*Averrhoa carambola*)
Slow-growing, very bushy, broad, tall (20 to 30 feet) tree with oblong, longitudinally angled fruit (2½ to 6 inches long and up to 3½ inches wide). The fruit has thin, orange-yellow skin and, when fully ripe, juicy, crisp, yellow flesh with flavor from very sour to mildly sweet. Can be eaten out of hand, sliced into salads or for garnish, as well as cooked in puddings, tarts, stews, and curries. Also known as star apple.

Bender, Gary S., and M. L. Arpaia. 1988. "Improved varieties" carambola: A specialty in the making? *Calif. Grower* 12(1):32, 38–39, 43.

Morton, J. F. 1987. *Fruits of warm climates,* 125–28. Winterville, N.C.: J. F. Morton.

Street, R. S. 1988. A star is born. *Calif. Farmer* 269(7):12–13, 22.

Sylvester, H. A. 1990. Growing carambolas in pots. *Fruit Gard.* 22(2):8, 15–17.

Ramirez, A. 1990. A star is born: Some California success stories. *Fruit Gard.* 22(2):8–9, 14–16.

CARDOON (*Cynara cardunculus*)
Easier to grow than its close relative the artichoke, this perennial's young stalks or older blanched stalks are cooked and used in salads or can be fried.

McDaniel, J. 1977. Cardoon, cardoni—whatever you call it, it's good. *Org. Gard.* 24(2):148–50.

CARISSA (*Carissa macrocarpa*)
Natal plum: Fast-growing shrubs up to 18 feet tall with red, plum-shaped 2-inch fruit that has a sweet, cranberry flavor.

Morton, J. F. 1987. *Fruits of warm climates,* 420–22. Winterville, N.C.: J. F. Morton.

Thomson, P. H. 1976. The carissa in California. *CRFG Yearbook* 8:73–81.

CAROB (*Ceratonia siliqua*)
A chocolate substitute. Evergreen tree (40 feet) with big, brown pods that are sweet and chewy. Can be eaten out of hand or milled for use as a chocolate substitute. Minimal water needs.

Morton, J. F. 1987. *Fruits of warm climates,* 121–24. Winterville, N.C.: J. F. Morton.

Thomson, P. H. 1971. The carob in California. *CRFG Yearbook* 3:61–102.

CELERIAC (*Apium graveolens* var. *rapaceum*)
Form of celery grown for its large, rounded roots, which are peeled, then used raw or cooked in salads.

Larkcom, Joy. 1984. *The salad garden: Salads from seed to table*, 93. New York: Viking Press. Very brief but adequate enough to experiment from.

Metcalf, H. N. 1972. Unusual vegetables. *Plant Gard.* 28(2):68–69.

CELTUCE (*Lactuca sativa* var. *asparagina*)
As the main harvest is the flower stalk, it can stand somewhat warmer temperatures than most lettuce.

Meeker, J. 1980. Celtuce: Lettuce plus. *Org. Gard.* 28(1):77–78.

Metcalf, H. N. 1972. Unusual vegetables. *Plant Gard.* 28(2):69–70.

CHAYOTE—See *Gourd: Chayote*

CHERIMOYA (*Annona cherimola*)
Semideciduous evergreen tree with 2-pound, heart-shaped fruit with creamy white, custard-textured flesh that has a pineapple-banana flavor. Must be hand pollinated.

Brown, T., and J. Brown. 1972. Cherimoyas in Carpeteria. *CRFG Yearbook* 4:2–6. Especially useful for propagation and transplanting.

Cherimoya section. 1983. *CFRG Yearbook* 15:5–43. The detail can be excruciating but is very thorough.

Englehart, Orton H. 1974. Pruning and pollinating the cherimoya. *CFRG Yearbook* 6:215–20. Best article on this particular aspect.

Gomez, M. C. 1983. The cherimoya. *CRFG Yearbook* 15:5–29. Although written for Spain, has many cultivation details not covered elsewhere.

Kadish, R. 1985. Cherimoya: A heretic's views. *Fruit Gard.* 17(1):5–9. Good cultural ideas with some interesting suggestions for simple experiments.

Morton, J. F. 1987. *Fruits of warm climates*, 65–69. Winterville, N.C.: J. F. Morton. Not as thorough as the above, but an alternative if they can't be found.

Neitzel, J. 1982. The cherimoya riddle. *CRFG Newsletter* 14(3):8–12. Deals mainly with pollination techniques but also has helpful information on irrigation and fertilization.

Rathore, D. S. 1990. Custard apple. In T. K. Bose and S. K. Mitra (eds.), *Fruits: Tropical and subtropical*, 449–68. Calcutta: Naya Prokash. Good added details.

Thomson, Paul H. 1970. The cherimoya in California. *CRFG Yearbook* 2:20–34. Article not very detailed but gives enough cultivation information to be successful.

CHESTNUT (*Castanea* spp.)
Jaynes, R. A. 1979. Chestnuts. *Nut tree culture in North America*, 111–27. Hamden, Conn.: North America Nut Growers Assoc.

Shah, A., S. S. Bora, and A. J. Roy. 1987. Effect of different mulchings and fungicidal treatments on germination of chestnut. *Prog. Hort.* 19(1/2):145–48.

Smith, A. H. 1976. The chestnut. *CRFG Yearbook* 8:15–51.

CHICORY

Belgian endive (*Cichorium endivia*)
Corey, K. A., et al. 1988. Witloof chicory: A new vegetable crop in the United States. In J. Janick and J. E. Simon (eds.), *Advances in new crops*, 414–18. Portland, Ore.: Timber Press. Production methods including suggestions for forcing box.

Hanauer, Gary. 1985. European chicory lessons. *Org. Gard.* 32(5):85–90. Fairly detailed account of how it's done in Dixon.

Endive
Benoit, F., and N. Ceustermans. 1985. Advancing the harvest of bolt-sensitive endive by means of temporary single and double crop covering. *Acta Hort.* 198:269–75.

Chaplin, L. 1991. Don't be bitter! Spice up your spring with great tasting endive! *Org. Gard.* 38(3):82–85.

Radicchio (*Cichorium intybus*)
Red- and green-leafed chicories with lettuce-like heads and, if the weather is cool enough, only a slight bitter flavor. Used raw in salads.

Cerme, M. 1990. Timing of sowing and harvesting of different chicory cultivars. *Acta Hort.* 267:69–76. Because the chicories can bolt and not head, depending on weather and cultural techniques, the presented climatic data can be combined with the "standard" cultural techniques.

Chandoha, Walter. 1984. Grow Italian greens: Radicchio, escarole and arugula. *Org. Gard.* 31(5):80–84.

CHIVES—See *Herbs: Allium – Chives*

CITRON (*Citrus medica*)
Slow-growing, 8- to 15-foot shrub or tree with fragrant, lemon-like fruit (from 3½ inches to 1 foot) that is firm, not very juicy, acid or sweet. The peel is candied.

Morton, J. F. 1987. *Fruits of warm climates*, 179–82. Winterville, N.C.: J. F. Morton.

CITRUS spp.
Handy citrus chart. 1988. *IC&RF Soc. Newsl.* 24 (Winter):10–11. Rootstock, edible fruit species.

Ray, R., and L. Walheim. 1980. *Citrus: How to select, grow and enjoy*, chaps. 2, 3, 4. Tucson, Ariz.: H-P Books. Detailed coverage of adaptable varieties of the 10 most common citrus and needed growing conditions.

CITRUS—JAPANESE SUMMER ORANGE (Pummelo × Mandarin natural hybrid)
Stands winter temperatures to around 20°F, with a December mean of 47°F and a January mean of 42°F. It has orange characteristics.

Yamamura, H. 1990. Natsudaidai: The Japanese summer orange. *Pomona* 23(4):23–24. Seeds obtainable for (U.S.) $2.00 from: Prof. Hiroshi Yamamura, Faculty of Agriculture, Shimane University-Matsue, Shimane 690, Japan.

CORN, SWEET (*Zea mays*)
Besides the vast number of sweet corn varieties, check at about two days after silk emergence to see if the variety is sweet enough to use as baby ("Asian") corn.

CORN SALAD (*Valerianella locusta*)
Very variable late-winter or early spring (depending on the variety) salad green with mild, nutty flavor.

Larkcom, Joy. 1984. *The salad garden: Salads from seed to table*, 81–82. New York: Viking Press.

CUCUMBER (*Cucumis sativa*)
Whealy, Kent. 1988. *The garden seed inventory.* Decorah, Ia.: Seed Saver Publications. For unusual nonhybrid varieties available in 1987. For hybrids and post-1987 nonhybrid varieties, browse general and "specialty" seed catalogs.

CURRANTS—See *Berries: Currants and gooseberries*

CURRY LEAF TREE (*Murraya koenigii*)
Spread across much of southern Asia and up to 5,000 feet in the Himalayas. The leaves are used like bay leaves and are an important flavoring agent in curries and chutneys.

Morton, J. F. 1984. The curryleaf tree is attracting attention in Florida. *Proc. Fla. State Hort. Soc.* 97:314–17.

Parmar, C., and J. Kumar. 1980. Studies on the sexual and vegetative propagation of *Murraya koenigii*. In P. K. Khosla (ed.), *Improvement of Forest Biomass*, 183–86. Solan, India: Indian Society of Tree Scientists.

DASHEEN—See *Taro: Dasheen*

EGGPLANT (*Solanum melongena* var. *esculentum*)
Whealy, Kent. 1988. *The garden seed inventory.* Decorah, Ia.: Seed Saver Publications. For unusual nonhybrid varieties available in 1987. For hybrids and post-1987 nonhybrid varieties, browse general and "specialty" seed catalogs.

EMBLIC (*Phyllanthus emblica*)
Graceful evergreen tree up to 60 feet tall with various-colored, hard fruit with thin, translucent skin adherent to the crisp, juicy, astringent, very acid flesh. It is used raw as a thirst-quencher, cooked in tarts, used as a seasoning, and made into jams, sweet meats, pickles, and relishes.

Bajpai, P. N., and H. S. Shukla. 1985. Aonla. In T. K. Bose (ed.), *Fruits of India: Tropical and Subtropical*, 591–600. Calcutta: Naya Prokash. The most complete cultivation instructions I've found.

Morton, J. F. 1987. *Fruits of warm climates*, 213–17. Winterville, N.C.: J. F. Morton.

Parmer, C. 1988. Himalayan wild amla may succeed in less cold southern states. *Pomona* 22(1):43–46. Could be a worthwhile $5 investment.

FAVA, BROAD BEAN—See *Beans: Fava, broad*

FEIJOA (*Feijoa sellowiana*)
Bushy, 3- to 20-foot shrub with fragrant, oblong fruit (1½ to 2½ inches long by 1½ to 2 inches wide) that has thick, white, granular, watery flesh with a flavor suggesting a combination of pineapple-guava or pineapple-strawberry. The fruit is eaten raw as a dessert or in salad or cooked in puddings, fritters, and so on.

Kirkpatrick, Shirley. 1988. Is feijoa the next kiwi fruit? *Calif. Grower* 12(2): 49–50. A few added details.

Morton, J. F. 1987. *Fruits of warm climates*, 367–70. Winterville, N.C.: J. F. Morton.

Rice, J. E. 1990. The pineapple guava: Hardier than you think. *Fruit Gard.* 22(4): 11. Survives 5°F winters to 105°F summers in North Carolina. Article also covers cultivation, albeit briefly.

Thomson, Paul H. 1970. Pineapple guava. *CRFG Newsletter* 2(2):2–4. With the preceding article, enough to get started.

FIGS (*Ficus carica*)
Arends, G. 1990. Three ways to prune a fig tree. *Fruit Gard.* 22(3):10–11. Gives a fairly detailed presentation of how to prune in relation to whether the fruit is for fresh, dried, or caprifig use.

Beutel, James. 1974. *Figs in the home garden*. University of California Cooperative Extension, Leaflet 2481.

Born, F. M. 1988. Figs. *Pomona* 21(2):20–30.

Condit, I. J. 1947. *The fig*. Waltham, Mass.: Chronica Botanica. This is apparently the latest work in English about the cultivation of figs and still considered the standard work.

Gerdts, Marvin, et. al. 1978. *Commercial dried fig production in California*. University of California Cooperative Extension, Leaflet 21051.

Mattern, V. 1991. Discover figs for your world! *Org. Gard.* 38(7):30–35. Grow figs organically.

FILBERTS—See *Nuts*

FLOWERS, EDIBLE
Carter, T. R. 1989. Rose hips. *IC&RF Soc. Newsl.* 28 (Winter):17.

Gordon, J. 1968. *The art of cooking with roses*, 159. New York: Noonday.

Gore, W. W. 1989. Guess what's growing for dinner? *Calif. Farmer* 270(5):54–57. Article lists a few more edible flowers and some recipes.

Greenfield, J. 1991. Growing and marketing edible flowers. *Bus. of Herb.* 9(1):6–9. Growing information is scarce, but some very good information on harvesting and marketing.

Harrington, G. 1985. Day lilies. *Grow your own Chinese vegetables*, 174–77. Pownal, Vt.: Garden Way Publishing.

Larkcom, Joy. 1984. *The salad garden: Salads from seed to table*, 125–29. New York: Viking Press.

Nobbs, E. J. 1979. The rose as a fruit. *CRFG Yearbook* 11:7–13.

Smith, L. W. 1973. *The forgotten art of flower cookery*. New York: Harper & Row. Primarily a recipe book but has a number of flowers not listed elsewhere.

FRUITS

General
Neitzel, J. 1980. Deciduous fruit varieties. *CRFG Yearbook* 12:20–40. Deciduous fruits adapted to L.A. and San Diego area with a list of nursery sources.

Riley, J. M. 1973. Growing rare fruit in Northern California. *CRFG Yearbook* 5:67–90. Has a good introduction and a fairly long list of fruits with appropriate brief comments.

Riley, J. M. 1981. Characteristics of rare fruit propagation from seed. *CRFG Yearbook* 13:38–46. Pages 40–46 contain a chart covering more than 100 rare fruits grown in California, listing seed storage life, ways to store and break dormancy, days to germinate, and tree hardiness. The preceding 38 pages are a useful review of both theoretical and practical aspects of propagating by seed.

Seed Germination
Janick, J., and J. Moore (eds.). 1975. *Advances in fruit breeding*. W. Lafayette, Ind.: Purdue University Press. Has more edible fruits than the following entry, but readers must look in the Seed Handling section of the Breeding Techniques part for each nut or fruit.

Schopmeyer, C. S. 1974. *Seeds of woody plants in the United States*. USDA Agricultural Handbook 450. Mainly concerned with trees in a forestry conservation context, but quite a few edible fruits and nuts are included.

Temperate (Pears, plums, and apples grow in this area)
IC&RF Soc. 1989. Propagating 20 rare plants. *IC&RF Soc. Newsl.* 28 (Winter):4–7. Gives illustrated propagation methods, track record on various exotics, and propagation chart.

Winter, P. 1986. High altitude finds: *Annona purperea*, *A. diversifolia* and *Pouteria viride*. *CRFG Newsletter* 18(1):25. Peggy Winter is probably the only seed source for these "tropicals" grown in Mexican high-pine country.

GARLIC (*Allium sativum*)
Poncavage, J. 1991. Grow great garlic! *Org. Gard.* 38 (8):30–35. Decent cultivation guide plus garlic greens, flowers, garlic types, sources.

Sims, W., T. M. Little, and R. E. Voss. 1976. *Growing garlic in California*. University of California Cooperative Extension, Leaflet 2948.

Chinese
It has a somewhat different flavor and is more tolerant of minimal care. Its cloves are also smaller and a little more difficult to separate. The preferred garlic in southern China and parts of Southeast Asia.

Elephant (*A. scorodoprasum*)
Much milder and much larger than true garlic.

Nichols, E. 1989. Growing elephant garlic. *Fine Gard.* 8:28–31. With seed sources and a sidebar on more technical points by R. E. Voss, UC Davis Extension vegetable specialist.

Rocambole (*A. sativum* var. *ophioscorodon*)
A mild garlic-flavored *allium* that also produces bulbils-bearing heads. The bulbils are used for planting.

Meeker, J. 1979. Rocambole: The gentle garlic you'll love. *Org. Gard.* 25(7):46, 48, 50, 52, 53. This article also lists sources of the bulbils.

GINKGO (*Ginkgo biloba*)
Careful: The flesh is very odoriferous and must be removed to get at the nut, which has traditionally been eaten on special occasions in China.

Millikan, D. F. 1979. Ginkgo. In R. A. Jaynes (ed.), *Nut tree culture in N. America*, 186–87. Hamden, Conn.: North America Nut Growers Assoc.

GOOSEBERRIES—See *Berries: Currants and gooseberries*

GOURD

Bitter melon (*Mimordica charantia*)
Climbing, glabrous, annual herbaceous vine whose young shoots and fruits are eaten. The fruit is oblong and cylindrical, to 10 inches in length, pointed at both ends, ribbed, warty, and definitely bitter but not unpleasantly so. Widely used in tropical Asian cuisines.

Johnson, H. 1985. *Bitter melon*. University of California Cooperative Extension, Leaflet 21399. For the backyard garden, but can be upscaled.

Chayote (*Sechium edule*)
A tiny squash plant with ½- to 1-foot, pear-shaped and sized, unevenly furrowed, alabaster- to dark-green-colored, nonsticking prickled fruit with pale, crisp, fine-textured flesh that has the taste and consistency of cucumber, zucchini, and kohlrabi. It is used like summer squash: stuffed, pureed, pickled, and in chutneys.

Knott, J. E., and J. R. Deanon (eds.). 1967. *Vegetable production in Southeast Asia*, 138, 143, 146, 157, 159. Los Baños: University of the Philippines College of Agriculture.

Rubatsky, Vincent, and Mas Yamaguchi. 1997. *World vegetables: Principles, production, and nutritive values* (2nd ed.). New York.: Chapman and Hall Publishing Co.

Chinese winter melon (*Benincasa hispida*)
Large (up to 12 inches long by 8 inches wide and weighing up to 25 pounds) oblong, green squash with white flesh tasting like slightly sweet zucchini.

Harrington, G. 1978. *Grow your own Chinese vegetables*, 38–42. Pownal, Vt.: Garden Way Publishing.

Morton, J. F. 1971. The wax gourd, a year-round Florida vegetable. *Fla. State Hort. Soc.* 84:104–9.

Sundararajan S., and C. R. Muthukrishnan. 1982. A high-yielding ashgourd. *Ind. Hort.* 27(3):15. Gives basic cultivation techniques and address for ordering seeds from India.

Luffah or sponge (*Luffa acutangula, L. cylindrica*)
Cox, L. 1982. The amazing luffa plant. *Small Farmer's Journal* 6(2):54–55.

Porterfield, W. W., Jr. 1955. Loofah: The sponge gourd. *Econ. Bot.* 9:211–23.

Pointed (*Trichosanthes dioica*)
Krishna-Prasad, V. S. R., and D. P. Singh. 1987. Effects of training in pointed gourd for growth and yield. *Prog. Hort.* 19 (1–2):47–49. A 1-meter-high trellis leads to significantly higher production.

Squash
Drowns, G. 1988. A world of squash. *Fine Gard.* 3:58–62. Not only cultivation but a thorough overview of the three *Curcurbita* species characteristics.

GRANADILLA—See *Passiflora* spp. (Passion Fruit)

GRAPES
James, F. 1985. Grape varieties for cooler areas of California. *CRFG Yearbook* 17:69–74.

GROUND CHERRY (*Physalis pubescens*)
Low, annual, trailing plants 1 foot tall by 3 to 4 feet wide with yellow-orange to yellow ⅝-inch fruit that drops from its husk when ripe and is used for preserves, jams, and pies.

Dremann, C. 1985. *Ground cherries, husk tomatoes, and tomatillos*, 1–2, 5–6, 9–10. Redwood City, Calif.: Redwood City Seed Company.

Wolff, X. Y. 1991. Species, cultivar and soil amendments influence fruit production of two *Physalis* species. *HortSci.* 26(12):1558–59. Fresh manures gave higher yields.

GRUMICHANA (*Eugenia dombeyi*)
A close relative of the Surinam cherry, this evergreen tree grows up to 50 feet tall and bears sweet, white-fleshed fruit 30 days after flowering. Problems include small size, large seeds, apical sepals. Fruit can be eaten out of hand or used in jellies, pies, and jam.

Joyner, G. 1988. Fruit of the month. *Trop. Fruit News* 22(6):54. Has a few details not covered in J. F. Morton.

Morton, J. F. 1987. *Fruits of warm climates*, 390–91. Winterville, N.C.: J. F. Morton.

GUAR—See *Beans: Guar*

GUAVA (*Psidium guajava*)
Fast-growing, open, semideciduous tree that grows to 30 feet, with round- to pear- shaped fruit (up to 1 pound and 3 inches in diameter) that has tough skin and a pulpy center with many small edible seeds and a distinctive sweet to highly acid flavor. Eaten out of hand and used in preserves, jellies, and butters.

Mitra, S. K., and T. K. Bose. 1990. Guava. In T. K. Bose and S. K. Mitra (eds.), *Fruits: Tropical and subtropical*, 280–303. Calcutta: Naya Prokash. Because this concerns India, not as helpful as the others but has many more details.

Morton, J. F. 1987. *Fruits of warm climates*, 356–63. Winterville, N.C.: J. F. Morton.

Sweet, C. 1991. A commercial crop for desert areas. *Calif. Grower* 15(7):34–35, 40. No planting instructions but good information on picking, storage, processing.

Brazilian or Castilian (*Psidium guineense*)
Relatively slow-growing 3- to 10-foot shrub or 23-foot tree with wide (⅜ to 1 inch), round or pear-shaped, yellow-skinned fruit with thick, pale, yellowish flesh that has an acid, resinous, slight strawberry flavor and is used for baking and preserves.

Morton, J. F. 1987. *Fruits of warm climates*, 365–66. Winterville, N.C.: J. F. Morton.

Chilean (*Ugni molinae*)
Riley, J. M. 1971. *Ugni molinae* (*Marytus ugni*): The chilean guava. *CRFG Newsletter* 3(2):1–3.

Riley, J. M. 1973. Growing rare fruits in Northern California. *CRFG Yearbook* 5:80.

Strawberry (*Psidium cattleianum*)
Open silver-green shrubs or small tree to 20 feet tall, with round to oval 1-inch fruit that has red to purple skin and flesh and many seeds embedded in firm, white, sweet-tart, tangy, resinous pulp. Has a tart cherry flavor and is used for jams and jellies.

Morton, J. F. 1987. *Fruits of warm climates*, 363–65. Winterville, N.C.: J. F. Morton.

HAZELNUTS—See *Nuts: Filberts*

HERBS—General
Use a good herb book and a lot of catalogs to learn of the vast array of herbs and their availability.

Bahr, M. 1991. Getting your herb plants off to a good start. *Bus. of Herb.* 9(1):29–32.

Bahr, M. 1991. Happy transplanting. *Bus. of Herb.* 9(2):28–33.

Barker, A. V. 1989. Liming soils for production of herbs. *Herb Spice Med. Pl. Dig.* 7(4):1–4. With list of herbs in relation to sensitivity to soil acidity. Very helpful.

Barker, A. V. 1990. Mulches for herbs. *Herb Spice Med. Pl. Dig.* 8(3):1–5. Covers organic, inorganic, and plastics with relative advantages, disadvantages. Very thorough.

Barker, A. V. 1986. Organic fertilizers for herbs. *Herb Spice Med. Pl. Dig.* 4(3):1–2, 4, 7.

Barker, A. V. 1987. Fertilizing perennial plantings of herbs organically with mulches. *Herb Spice Med. Pl. Dig.* 5(2):3–4. Both are fairly general but have some good practical tips.

Corey, K. A. 1989. Postharvest preservation of fresh herbs: Fundamentals and prospects. *Herb Spice Med. Pl. Dig.* 7(3):1–5. Good presentation of different possibilities.

DeBaggio, T. 1987. Profitable annual herbs in less than 35 days. *Herb Spice Med Pl. Dig.* 5(3):7–8.

DeBaggio, T. 1987. Fast-cropping perennial potted herbs from seed: As easy as annuals but some cautions apply. *Herb Spice Med. Pl. Dig.* 5(4):5–6. Both articles present very useful information on rapidly growing good potted herbs.

DeBaggio, T. 1987. Many herbs tolerate more soil acidity than commonly believed. *Herb Spice Med. Pl. Dig.* (4):7–8.

Galambosi, B. 1991. Harvesting and cleaning herb seeds. *Herb Spice Med. Pl. Dig.* 9 (3):1–4. Good general practical information correlated with machine harvesting, drying.

Joyce, D., and M. Reid. 1986. Postharvest handling of fresh culinary herbs. *Herb Spice Med. Pl. Dig.* 4(2):1–2, 5–7. Thorough coverage of 10 herbs of different physical characteristics.

Kowalchik, C., and W. H. Hylton (eds.), *Rodale's Illustrated Encyclopedia of Herbs*, 262–67. Emmaus, Penn.: Rodale Press. Contains charts of herb pests and diseases with organic controls.

Long, J. 1992. Fresh herbs as a sideline. *Bus. of Herbs* 9(6):1–2, 31. Although the author supplements his dry herb business with fresh herbs, his points are applicable to most operations with a fresh herbs "sideline."

Miller, R. 1991. The farm: Some important considerations for the beginning herb farmer. *Herb Mark. Rep.* 7(1):1–4.

Ricotta, J. A., and J. B. Masiumas. 1991. The effects of black plastic mulch and weed control strategies on herb yield. *HortSci.* 26:539–49. Designed for small growers. Mulch on basil and rosemary with either herbicides or hoeing resulted in greater yields. Parsley results were inconclusive.

Schumann, G. 1989. Preventing diseases in culinary herbs. *Herb Spice Med. Pl. Dig.* 7(1):1–4, 10–11. Because of limited number of permitted chemical controls, emphasizes cultural, varietal practices.

Simon, J. E., A. F. Chadwick, and L. E. Craker. 1984. *Herbs: An indexed bibliography, 1971–1980: The scientific literature on selected herbs, and aromatic and medicinal plants of the temperate zone.* Hamden, Conn.: Archon Books. If you're willing to wade through many references, it's possible to find some very useful articles.

Simon, J. E., and D. R. Decocteau. 1983. The propagation of selected herbs. *Herb Spice Med. Pl. Dig.* 1(1):3–4. After a brief discussion of propagation methods, lists 61 herbs with recommended techniques.

Allium: Chives (*Allium schoenoprasum*)
Jones, H. A., and L. K. Mann. 1963. *Onions and their allies: Botany, cultivation, and utilization*, 230-33. New York: Interscience Publishers.

Allium: Chinese chives (*A. tuberosum*)
Garlic-flavored chive with white flowers smelling like roses.

Harrington, G. 1978. *Grow your own Chinese vegetables*, 135–40, 228–30. Pownal, Vt.: Garden Way Publishing.

Allium: Shallots (*A. cepa* var. *aggregatum*)
Jones, H. A., and L. K. Mann. *Onions and their allies: Botany, cultivation, and utilization*, 136–37. New York: Interscience Publishers.

Angelica (*Angelica archangelica*)
A tall, biennial herb needing some shade and moist, rich soil. The tender stems and petioles are candied, and the leaves are cooked like a vegetable.

Halva, S. 1990. Angelica: Plant from the North. *Herb Spice Med. Pl. Dig.* 8(1):1–3. Good instructions.

Laufer, G. A. 1984. The effect of stratification of *Angelica archangelica* seed after storage. *Plant Propag.* 30(2):13–15. Because the seed is almost impossible to germinate after storage, this stratification procedure is very important.

Miller, R. A. 1987. The cultivation and harvest of *Angelica* root. *Herb Mark. Rep.* 3(7):1–5.

Basil (*Ocimum basilicum*)
Hampstead, Marilyn. 1984. *The basil book.* New York: Pocket Books. Detailed account of varieties, cultivation, and uses.

Simon, Dr. J. E., and Dr. L. E. Craker. 1984. Introduction to sweet basil culture. *Herb Spice Med. Pl. Dig.* 2(2):1–2, 6. Good basic information.

Simon, J. E., and D. Reiss-Bubenheim. 1988. Field performances of American basil varieties. *Herb Spice Med. Pl. Dig.* 6(1):1–4. Both variety and seed company comparisons.

Capers (*Capparis spinosa*)
Foster, G. B., and R. F. Louden. 1980. Caper bush. *Park's Success with Herbs*, 61. Greenwood, S.C.: G. W. Park Seed Co.

Kontaxis, D. G. 1989. Capers: A new crop for California? *Small Farm News* (July-Aug.):3.

Catnip (*Nepeta cataria*)
Ferguson, J. M., W. W. Weeks, and W. T. Fike. 1988. Catnip production in North Carolina. *Herb Spice Med. Pl. Dig.* 6(4):1–4. Good presentation of small-scale commercial production with warning about ease of overproduction.

Cumin (*Cumin cyminum*)
Bhati, D. S., et al. 1986. Cumin: A tropical condiment. *Ind. Hort.* 31(3):27–28. Detailed instructions, but sowing time doesn't apply here, as the main requirement seems to be dry weather from flowering onward.

Fennel (*Foeniculum vulgare* var. *azoricum*)
Morales, M., D. Charles, and J. Simon. 1991. Cultivation of Finnochio fennel. *Herb Spice Med. Pl. Dig.* 9(1):1–4. Good information on the "bulb" fennel with comparison of 16 cultivars.

Simon, J. A. 1989. Fennel: New specialty vegetable. *Herb Spice Med. Pl. Dig.* (4):5.

Fenugreek (*Trigonella foenum-graecum*)
This legume is used as a forage and vegetable, and its seeds are used in curries and sprouted for salads. The seed extract is used for artificial maple syrup. Well suited in rotation to help make clay soils more friable.

Miller, R. A. 1987. Current and potential markets for fenugreek seeds. *Herb Mark. Rep.* 3(11):1–3.

Ginger (*Zingiber officinale*)
Cude, Kaye. 1985. For a lively spice—grow ginger. *CRFG Newsletter* 17(3):10–15.

Ginseng (*Panax quinquefolius*)
Duke, J. A. 1989. *Ginseng: A concise handbook*, 170–220. Algonac, Mich.: Reference Publications. The two chapters cover the cultivation and economics of growing ginseng.

Hawkins, A. 1988. Wild-simulated ginseng cultivation. *Bus. of Herbs* 5(6):1–2, 4, 27–29.

Persons, W. S. 1986. *American ginseng: Green gold*. Pompano Beach, Fla.: Exposition Press. A very complete book on the various ways of raising ginseng, complete with seed sources, costs, markets, etc.

Senteney, R. L. 1990. Growing and marketing ginseng. *Bus. of Herbs* 8(4):4–5, 29–33. Detailed report on the author's methods.

Williams, L., and J. Duke. 1978. *Growing ginseng*. USDA Farmers' Bulletin 2201.

Yang, Y. Y. 1974. *The effects of different shading of mulching on yield of root and quality in Panax ginseng*. Proc. Intl. Ginseng Symp., Cent. Res. Inst., Off. Monopoly, Rep. Korea, pp. 137–46. Straw was better than plastics in Taiwan.

Horseradish

Lathrop, N. J. 1981. *Herbs: How to select, grow, and enjoy*, 74–75. Tucson, Ariz.: H-P Books. Very basic, but easy to build from.

Lavender (*Lavendula* spp.)

Can be used fresh as an edible flower, dried in sachets, or distilled for its oil.

Miller, R. A. 1987. The cultivation and harvest of lavender. *Herb Mark. Rep.* 3(1):6–7.

Lemongrass (*Cymbopogon* spp.)

Long (2 inches) bunch grass with a combined lemon-lime peel and fresh hay flavor. The coarse outer stalks are used to flavor soups, and the pale inner stalks are finely slivered for use in a wide variety of Southeast Asian and West African cuisines.

Ghosh, M. L., and S. K. Chatterjee. 1977. Cultivation of *Cymbopogon* spp. in Burdwan. *Advances in the essential oil industry*, 29–36. New Delhi, India.

Miller, R. A. 1989. The cultivation and harvest of lemon grass. *Herb Mark. Rep.* 5(10):1–4.

Prasad, L. K., and S. K. Mukherji. 1980. Effect of nitrogen, phosphorus, and potassium on lemon grass. *Indian J. Agron.* 25(1):42–44.

Lovage (*Levisticum officinale*)

This celery-flavored perennial herb's seed and young stems are used as condiments and its blanched leafstalks and leaves as vegetables.

McDaniel, J. 1975. If you like celery, you'll love lovage. *Org. Gard. Farming* 22(3): 148–49. General, but has a few good tips.

Mint, licorice (*Agastache foeniculum*)

Miller, R. A. 1989. Licorice mint: A new herb tea ingredient. *Herb Mark. Rep.* 5(11):6–7.

Mint, orange bergamot (*Mentha citrata*)

Miller, R. A. 1989. The cultivation and harvest of orange mint. *Herb Mark. Rep.* 5(7): 1–4.

Parsley (*Petroselinum crispum*), Hamburg parsley (*P. c.* var. *tuberosum*)

The root of Hamburg or turnip-rooted parsley is used as a vegetable. Plain-leaved or Italian parsley has become popular in specialty restaurants and markets.

Miller, R. A. 1987. The cultivation and harvest of parsley. *Herb Mark. Rep.* 3(4): 1–5.

Simon, Dr. J. E., and M. L. Overley. 1986. A comparative evaluation of parsley cultivars. *Herb Spice Med. Pl. Dig.* 4(1):3–4, 7. Compares cultivars with each other and the same cultivars from different seed companies.

Rosemary (*Rosemarinus officinalis*)

DeBaggio, T. 1988. Growing rosemary. *Fine Gard.* 2:51–55. Covers varieties and their uses, sources, planting, and care pointers.

Saffron (*Crocus sativus*)

One of the most expensive spices, it is obtained by plucking and drying the stigmas of newly opened flowers.

Goss, C., and K. Skrovrinskie. 1989. Just wild about saffron. *Herb Quarterly* 43:35–38.

Mandan, C. L., B. M. Kapur, and U. S. Gupta. 1966. Saffron. *Econ. Bot.* 20:377–85. Covers large-scale planting techniques in a number of regions.

Sampathu, S. R., et al. 1984. Saffron: Cultivation, processing, chemistry and standardization. *CRC Crit. Rev. Food Sci. Nutr.* 20(2):126–32.

Sesame seed (*Sesamum indica*)

Miller, R. 1989. The cultivation and harvest of sesame seed. *Herb Mark. Rep.* 5(2):5–9.

Sweet cicely (*Myrrhis odorata*)
DeBaggio, T. 1988. Sweet cicely: Seed germination tricks. *Herb Spice Med. Pl. Dig.* 6(1):6. Ways to germinate a notoriously difficult seed.

Tarragon (*Artemisia dracunculus*)
VanHevelingan, A. 1990. French tarragon. *Fine Gard.* 11:25–27. Complete with cutting sources and a sidebar about growing through the winter.

HORSERADISH TREE (*Moringa pterygosperma*)
Evergreen tree, 25 feet or taller, with all parts except stem edible. The roots taste like horseradish, the leaves like mustard, the green pods somewhat like beans, and the roasted mature pods somewhat like peanuts. Caution: Seeds contain saponin so should not be eaten in large quantities.

Bond, R. E. 1985. The horseradish tree. *CRFG Newsletter* 17(4):14–16.

Ramachandran, C., et al. 1980. Drumstick: A multipurpose Indian vegetable. *Econ. Bot.* 34:276–83. Indian propagation methods, spacing, care.

Sharma, G. K., and V. Raina. 1980. Propagation techniques of *Moringa oleifera*. In P. K. Khosla (ed.), *Improvement of forest biomass*, 175–81. Solan, India: Indian Society of Tree Scientists.

ICE CREAM BEAN (*Inga affinis*)
The seeds of this legume are surrounded by sweet pulp that tastes like vanilla ice cream.

Duke, J. A. 1981. *Handbook of legumes of world economic importance*, 101–2. New York: Plenum Press.

IC&RF Soc. 1987. 10-page illustrated rare fruit chart. *IC&RF Soc. Newsl.* 23:11. Little information but does list two nurseries that carry the bean.

JABOTICABAS (*Myrciaria cauliflora*)
A slow-growing, bushy, evergreen tree up to 30 feet tall with reddish-purple, grape-size, trunk-borne fruit that has gelatinous grape-flavored pulp that is used in jams and wine.

Morton, J. F. 1987. *Fruits of warm climates*, 371–74. Winterville, N.C.: J. F. Morton.

Parmar, C., and M. K. Kaushal. 1982. *Wild fruits of the sub-Himalayan region*. Ludhiana, India: Kalyani Pub. See below for purchase details. Dr. Chiranjit Parmar, in a letter to the editor in the June 1990 *Fruit Gardener* magazine, offers to send a copy of the book via registered air mail for $15 (U.S.) in check (not cash) form. In addition, he also will help interested readers of the book obtain seed. His address: Dr. Chiranjit Parmar, Sharda Niwas (Near Ceeta Ashram), Thodo Ground, Solan, H.P. 173 212 India. Based on visits to California, Dr. Parmar states that the climate of the Himalayan mid-hills is somewhat similar to that of parts of California. The book covers in fair detail 26 wild fruits and includes judgments on their quality and potential uses.

Wiltbank, W. J., et al. 1983. The jaboticaba in Brazil. *Proc. Am. Soc. Hort. Trop. Reg.* 27:57–69. Has some good propagation tips (p. 64) and an extensive list of cultivars (p. 63).

JACK FRUIT AND RELATED SPECIES (*Artocarpus* spp.)
Evergreen tree that grows to 65 feet with large (up to 80 pounds) trunk-borne fruit with soft, juicy flesh and large seeds that taste like chestnuts when roasted.

Morton, J. F. 1987. *Fruits of warm climates*, 58–64. Winterville, N.C.: J. F. Morton.

JAMBOLAN (*Syzygium cumini*)
Fast-growing, wide-spreading 40- to 50-foot tree with clusters of nearly black ½- to 2-inch round or curved fruits with thin, smooth, glossy skin. The fruit has very juicy purple or white astringent flesh that is acid to fairly sweet and is eaten raw or made into tarts, sauces, jams, and vinegar.

Bajpai, P. N., and O. P. Chaturvedi. 1990. Jamon. In T. K. Bose and S. K. Mitra (eds.), *Fruits: Tropical and subtropical*, 747–56. Calcutta, India: Naya Prokash. Much more detailed than J. F. Morton.

Morton, J. F. 1987. *Fruits of warm climates*, 375–78. Winterville, N.C.: J. F. Morton.

JERUSALEM ARTICHOKE (*Helianthus tuberosus*)

A member of the sunflower family, this easily grown vegetable produces tubers that, unlike potatoes, can be eaten by diabetics because they are nearly starchless and contain insulin. With a sweet, nutty flavor, they can be mashed, stir-fried, pickled, made into soup, or chilled and sliced into salads.

Boswell, V. R., et al. 1936. *Studies of the culture and certain varieties of the Jerusalem artichoke.* USDA Technical Bulletin 514. The whole bulletin is useful, particularly pp. 68–69, but the "Brief Cultural Recommendations" at the end is sufficient.

Dallimonti, L. 1979. The alternative potato. *Org. Gard.* 24(6):34, 36. Good growing and variety suggestions.

Sebert, E. 1974. Getting started with Jerusalem artichokes. *Org. Gard.* 21(4): 66–67. Additional useful growing suggestions.

Wilson, C. M. 1976. They "eat best" in late winter. *Org. Gard.* 23(8):62–63. Some suggestions for digging and storing.

JOJOBA (*Simmondsia chinensis*)

Very branchy, deciduous 6-foot shrub of which the females bear edible, ¾-inch-, acorn-like fruit with slightly filbert-like flavor. Fruit also yields oil used in shampoos and as a sperm oil substitute.

Thomson, P. H. 1975. Jojoba horticulture. *CRFG Yearbook* 7:98–162.

Yermanos, D. M. 1974. Jojoba: A crop whose time has come. *Calif. Agric.* 33(7/8):4–7, 10–11.

JUJUBE (*Ziziphus jujuba*)

Upright, deciduous, 30-foot evergreen shrub or small tree with egg-shaped, 2-inch- long fruit with hard-shelled stones and crisp, sweet white flesh. Can be eaten out of hand or used in preserves, syrup, jelly, puddings, and cheese.

Hagar, E. T. 1988. My favorite fruit tree: The jujube. *Fruit Gard.* 21(3):13–15.

Lee, B.W. 1960. *The jujube or "Chinese date."* University of California Cooperative Extension, Leaflet 2729.

Meyer, R. 1991. Growing jujube from seed. *Pomona* 24(1):61–62.

Parrek, O. P. 1983. *The Ber.* New Delhi: ICAR. Although it deals with *Z. mauritiana*, cultivation is the same and is presented in much more detail.

Riley, J. M. 1969. The Chinese jujube. *CRFG Yearbook* 1:23–29. Most thorough account I've found.

Sharma, V. P., and V. N. Kore. 1990. Ber. In T. K. Bose and S. K. Mitra (eds.), *Fruits: Tropical and subtropical*, 592–615. Calcutta: Naya Prokash.

Zhongning, H., and Z. Xiang. 1991. Effects of foliar spray of urea and ethephon in autumn on nitrogen storage and reutilization in jujube trees. *Acta Hort. Sinica* 18 (2):106, 105, 104. Ethephon is a 21.4 percent water-soluble phosphorous fertilizer. Some brand names are Florel, Ethrel, Cepha.

KARANDA (*Carissa congesta*)

Rank-growing, straggly, climbing shrub, 10 to 15 feet tall, with nearly black, skinny, ½- to 1-inch fruit in clusters of 3 to 10 with juicy, red or pink, acid to fairly sweet pulp. The fruit can be eaten out of hand, stewed, juiced, or used in curries, tarts, puddings, and chutney.

Morton, J. F. 1987. *Fruits of warm climates*, 422–24. Winterville, N.C.: J. F. Morton.

KEI APPLE (*Dovyallis caffra*)

Will survive brief temperatures to 20°F. Often spiny, 30-foot-tall by 25-foot-wide shrub or small tree with nearly round, bright-yellow, somewhat tough-skinned aromatic fruit with mealy, apricot-textured, highly acid, juicy flesh eaten as a dessert sprinkled with sugar or used in jams and jellies.

Morton J. F. 1987. *Fruits of warm climates*, 315–17. Winterville, N.C.: J. F. Morton.

Silber, D. 1989. Update on Kei apple. *IC&RF Soc. Newsl.* 28:19.

KERIBERRY

Very vigorous, heavily armed, thick-stemmed, blackberry-raspberry hybrid with a pronounced but mild blackberry flavor. A late-winter or early spring bearer. A New Zealand berry with apparently good prospects for California.

Nobbs, K. J. 1986. The keriberry. *CRFG Newsletter* 18(2):10–11. Also has address for getting plants.

KIWANO, AFRICAN HORNED MELON (*Cucumis metuliferus*)

Although considered a weed in parts of the Central Valley, the jelly-like interior of the fruit is considered a delicacy among others.

Benzioni, A., S. Mendlinger, and M. Ventura. 1991. Effect of sowing dates, temperatures on germination, flowering and yield of *Cucumis metuliferus*. *HortSci.* 26(8):1051–53.

KIWIFRUIT (*Actinidia argula*)
Common (*A. deliciosa*)

Beutel, J. A. 1988. Kiwifruit. In J. Janick and J. E. Simon (eds.), *Advances in New Crops*, 309–16. Portland, Ore.: Timber Press. Includes production costs, problems of oversupply.

Family Farm Series. 1989. *Kiwifruit production in California*. Small Farm Center, University of California, Davis, p. 7.

KOHLRABI (*Brassica oleracea* var. *gongylodes*)

Larkcom, Joy. 1984. *The salad garden: Salads from seed to table*, 88-90. New York: Viking Press.

KUMQUAT (*Fortunella crassifolia, F. margarita, F. japonica*)

Slow-growing shrubby, almost spineless, 8- to 15-foot tree with oval or round, yellow or orange fruit (⅝ to 1½ inches wide) with little acid or subacid pulp. The whole fruit is eaten raw or preserved, candied, pickled whole, and used for making marmalade.

Hodgson, R. W. 1967. Horticultural varieties. In W. Reuther, et al. (eds.). *The citrus industry*, 580–84. Berkeley: University of California Division of Agricultural Sciences.

Morton, J. F. 1987. *Fruits of warm climates*, 182–85. Winterville, N.C.: J. F. Morton.

LAMB'S LETTUCE, MACHE, CORN SALAD (*Valerianella olitoria*)

Traditionally, the earliest spring salad green.

Corbineau, F., and D. Come. 1990. Effects of priming on the germination of *Valerianella olitoria* seeds in relation with temperature and oxygen. *Acta Hort.* 267:191–97. Because the germination of lamb's lettuce seed is very sensitive to too high (+75°F) or too low (–40°F) temperatures, a method of priming the seeds to help germination at the extreme temperatures is presented.

LEEKS (*Allium ampeloprasum*)

Jones H. A., and L. K. Mann. 1963. *Onions and their allies: Botany, cultivation, and utilization*, 233–37. New York: Interscience Publishers.

Mallon, K. 1991. Plugging leeks. *Nat. Gard.* 14(6):50–51. Involves layering bed with newspaper and straw and dibbling in leeks. Possible to use plastic instead of newspaper.

LEMON, GALGAL (*Citrus pseudolimon*)

Hardier than the apple, galgal is a vigorous, 15- to 20-foot upright tree with ½- to 1-pound yellow fruit with coarse, pale yellow, moderately juicy, very sour flesh that is used much like the lemon. To obtain seeds for (U.S.) $3, which covers handling and mailing costs, contact Dr. Chiranjit Parmar, Sharda Niwas (Near Ceeta Ashram), Thodo Ground, Solan, H.P. 173 212 INDIA.

Parmar, C. 1990. Galgal: A cold hardy lemon from India. *Pomona* 23(3):78–79.

LETTUCE (*Lactuca sativa* var. *crispus, capitata, longifolia*)

Whealy, K. 1988. *The garden seed inventory*. Decorah, Ia.: Seed Saver Publications. For unusual nonhybrid varieties available in 1987. For hybrids and post-1987 nonhybrid varieties, browse general and "specialty" seed catalogs.

LIME

Mandarin

Stands cold better than the lemon. All are fast-growing bushy trees or wide-spreading bushes.

Morton, J. F. 1987. *Fruits of warm climates*, 178–79. Winterville, N.C.: J. F. Morton.

Mexican (*Citrus aurantifolia*)
The bartender's lime grows on vigorous, shrubby, spiny trees that are from 6½ to 13 feet tall. Besides its use in alcoholic drinks, the juice of the 1- to 2-inch-diameter, green to pale-yellow fruit is used to garnish fish and meats and to make syrup, sauces, and pie. Whole fruits are used in jam, jelly, and marmalade and are pickled.

Hodgson, R. W. 1967. Horticultural varieties. In W. Reuther, et al. (eds.), *The citrus industry*, 575–76. Berkeley: University of California Division of Agricultural Sciences.

Morton, J. F. 1987. *Fruits of warm climates*, 168–72. Winterville, N.C.: J. F. Morton.

Sweet (*C. limettiodes*)
Nearly thornless, moderately vigorous, 15- to 20-foot tree with greenish to orange-yellow, nearly round fruit that has very juicy, nonacid, bland, fairly bitter pulp. Can be eaten out of hand or cooked in preserves.

Hodgson, R. W. 1967. Horticultural varieties. In W. Reuther, et al. (eds.), *The citrus industry*, 578–79. Berkeley: University of California Division of Agricultural Sciences.

Morton, J. F. 1987. *Fruits of warm climates*, 175–76. Winterville, N.C.: J. F. Morton.

Tahiti (*C. latifolia*)
Tree is similar to the sweet lime, with pale-yellow fruit (1½ to 2½ inches wide by 2 to 3 inches long) that has a greenish-yellow, tender, acid pulp and is used like the Mexican lime.

Hodgson, R. W. 1967. Horticultural varieties. In W. Reuther, et al. (eds.), *The citrus industry*, 576–77. Berkeley: University of California Division of Agricultural Sciences.

Morton, J. F. 1987. *Fruits of warm climates*, 172–75. Winterville, N.C.: J. F. Morton.

LITCHI (*Litchi chinensis*)
Tall (30 to 100 feet) and broad, slow-growing evergreen tree with clusters of red aromatic 1- by 1½-inch fruits that have glossy, succulent, subacid flesh. The fruit can be eaten out of hand, stuffed, pureed, canned, pickled, and used as a topping on ham and steak.

Langdon, R. D. 1969. The lychee. *CRFG Yearbook* 1:1–8.

Louscher, R. 1980. The *Sapindaceae* family. *CFRG Yearbook* 12:41–45.

Maity, S. C., and S. K. Mitra. 1990. Litchi. In T. K. Bose and S. K. Mitra (eds.), *Fruits: Tropical and subtropical*, 420–48. Calcutta: Naya Prokash. More details that can be used in conjunction with the other references.

Menzel, C. 1990. Nitrogen: The key to successful lychee production. *Calif. Grower* 14(7):28–30. Quite useful discussion and guidelines on this aspect of litchi production.

Morton, J. F. 1987. *Fruits of warm climates*, 249–59. Winterville, N.C.: J. F. Morton.

Sauco, V. G., and U. G. Menini. 1989. *Litchi cultivation*. FAO Plant Production and Protection Paper 83, Rome, Italy, pp. 39–73. A detailed exposition that, although it attempts worldwide coverage, has information that is useful for California.

LONGAN (*Euphoria longana* or *Dimocarpus longan*)
Tall (30 to 40 feet) and wide (45 feet) evergreen tree with clusters of ½- to 1-inch round, thin-skinned, brownish fruit with flesh somewhat like the litchi. It can be eaten out of hand, cooked, canned, dried, or used in liqueurs.

Menzel, C. M., et al. 1990. Longan. In T. K. Bose and S. K. Mitra (eds.), *Fruits: Tropical and subtropical*, 522–46. Calcutta: Naya Prokash. Much greater detail, especially concerning climate.

Morton, J. F. 1987. *Fruits of warm climates*, 259–61. Winterville, N.C.: J. F. Morton.

LOQUAT (*Eriobotrya japonica*)
An evergreen 20- to 30-foot-tall tree with clusters of oval, 1- to 2-inch, yellow- to orange-skinned fruits with succulent, sweet to subacid pulp. Eaten fresh, stewed, and as a sauce, made into jam and jelly, canned, and spiced.

Jerris, W. 1990. Evaluation of cultivars. *IC&RF Soc. Newsl.* 33:14.

Loquat section. 1985. *CRFG Yearbook* 17:1–39.

Morton, J. F. 1987. Loquat. *Fruits of warm climates*, 103–8. Winterville, N.C.: J. F. Morton.

Pathak, R. K., and H. D. Gautam. 1985. In T. K. Bose (ed.), *Fruits of India: Tropical and sub-tropical*, 549–58. Calcutta: Naya Prokash.

Sweet, C. 1986. Asian-Americans favor fresh loquat: Part I. *Calif. Grower* 10(2): 12, 30.

Sweet, C. 1986. Loquat: Part 2. *Calif. Grower* 10(4):30, 32. Discusses possible pollination problems with some varieties.

LOTUS, INDIAN or CHINESE (*Nelumbo nucifera*)
The tubers, seeds, leaves, and leaf stalks are used in Asian and Native American cooking.

Creasy, R. 1982. *The complete book of edible landscaping*, 260–62. San Francisco: Sierra Club.

Perry, F. 1938. *Water gardening*, 77–78. London: Country Life.

LUFFA or SPONGE GOURD—See *Gourd: Luffa or sponge*

MACADAMIA NUT—See *Nuts: Macadamia nut*

MANDALO
A mandarin × pummelo hybrid that is both prolific and precocious. Its fruit holds on the tree very well and is very juicy with a unique, somewhat watermelon-like flavor. Because of its seediness, it is either juiced or eaten like grapefruit. Starts bearing from seed in two to three years in Riverside. Has a thin rind, so it needs careful handling and protection from prolonged freezing. The Indoor Citrus and Rare Fruit Society would like it to be tried in the cooler climates to check its adaptability.

Moore, P. W. 1987. Announcing the new citrus hybrid "mandalo." *IC&RF Soc. Newsl.* 21:10–12.

MANDARIN ORANGE
IC&RF Soc. 1988. The many mandarins (also tangelos and tangers). *IC&RF Soc. Newsl.* 27:10–12. Presents a chart of varieties, regional adaptability, fruiting time, etc.

MANGO (*Mangifera indica*)
Dense, evergreen, 70-foot tree with smooth, thin-skinned, egg-shaped fruit (up to 1½ pounds) with fibrous, light-orange, sweet, peach-like flavored flesh. Eaten out of hand or used in chutneys and sherbets.

Rajan, S., and G. C. Sinha. 1987. Use of aluminum foil for increasing veneer grafting success under adverse conditions. *Prof. Hort.* 19(1/2):141–42. Allows grafting over an additional five months.

Staedeli, J. H. 1972. Establishing a mango grove in San Diego, California. *CRFG Yearbook* 4:11–18.

Staedli, J. H. 1991. A commercial crop in avocado country. *Calif. Grower* 15(9): 33–35. Good information on trial practices, varieties, yields.

Thomson, P. H. 1969. Growing mangos in Southern California. *CRFG Yearbook* 1:9–22.

MAYHAW (*Crataegus opaca*)
Heavy bearing, deciduous, quite variably sized tree producing ¾-inch diameter, apple-like fruit in April and May. Fruit can be eaten out of hand or, as is more usual, made into a unique-tasting, delicious jelly.

Akin, S. 1985. Mayhaws in California. *CRFG Newsletter* 17(1):15–17.

Akin, S. 1990. Mayhaw: An edible hawthorn. *IC&RF Soc. Newsl.* 32:12–13.

Moore, B. 1991. Special procedures for raising mayhaw species from seed. *Pomona* 24(1):28–29.

Moore, B. 1991. Mayhaw tree-cleft grafting: Summer set method. *Pomona* 24(1):18–19. With parafilm tape, cleft grafting can be done throughout the summer.

Payne, J., and G. Krewer. 1990. Mayhaw: A new fruit crop for the South. In J. Janick and J. E. Simon (eds.), *Advances in new crops*, 317–20. Portland, Ore.: Timber Press.

MEADOWFOAM (*Limnanthes alba*)
A good winter oilseed crop, especially for poorly drained soils, that can be followed by late-planted summer vegetable crops. Same growing and harvesting equipment as for grass seed produc-

tion. The steamed meal can partially replace soybean meal in rabbit and chicken rations. Raw meal can replace alfalfa for nongestating sheep and cattle. In low rainfall years, should be irrigated at the onset of flowering.

Miller, R. A. 1987. Meadowfoam: A potential oilseed crop. *Herb Mark. Rep.* 3(10):1–3.

MEDLAR (*Mespilus germanica*)
A slow-growing, 10- to 20-foot tree with 1- to 2-inch apple- to pear-shaped fruit that is either allowed to fall from tree or picked after first frost and is near decay (the term for this is *bletting*), soft, and mellow. Can be eaten out of hand or, at the "proper" stage, made into a jelly.

Baird, J. R., and J. W. Thieret. 1989. The medlar from antiquity to obscurity. *Econ. Bot.* 43:333–38, 364.

Polkowski, G. R. 1976. The medlar. *CRFG Yearbook* 8:119–20.

MELONS (*Cucumis melo, Citrullus vulgaris*)
For information on minimelons, seedless or hybrid triploid watermelons, and cantaloupes, send for *A small-scale agricultural alternative: Dessert vines*, 1989, available free of charge from Bud Kerr, USDA-CSTS-OSSA, Suite 242-D, Washington, D.C. 20251. Also check seed catalogs for others.

MONSTERA DELICIOSA (Split-Leaf Philodendron)
Houseplant everywhere. Evergreen, strong, woody climber with 10-inch-long cone-like fruit that, when scales fall and the cone turns yellow, has a banana-pineapple-flavored pulp that is eaten out of hand as a dessert or used for preserves.

Fisch, B. E. 1976. *Monstera deliciosa*. *CRFG Yearbook* 8:89.

Morton, J. F. 1987. Ceriman. *Fruits of warm climates*, 15–17. Winterville, N.C.: J. F. Morton.

Tamblyn, G. 1976. Flowering and fruiting of *Monstera deliciosa*. *CRFG Yearbook* 8:90.

For nursery sources see *IC&RF Soc. Newsl.* 23:12 (1987).

MULBERRY, BLACK (*Morus nigra*)
The 10- to 30-foot trees produce very soft, sweet, blackberry-like fruit that in Europe is eaten out of hand and used as a dessert and for wines and syrups. Caution: Birds love them.

Fraser, S. 1924. *American fruits*, 582–85. Orange Judd, New York. This brief article covers propagation, grafting, planting, and harvesting.

Fraser, S. 1989. What they say about the mulberry. *IC&RF Soc. Newsl.* 30:14–15. Includes nursery sources.

MUNG BEANS —See *Beans: Mung*

MUSHROOMS
Kurtzman, R. H. 1978. A vertical tray system for the cultivation of edible fungi. *Mushroom Sci.* 10(2):429–36. Detailed instructions for building vertical trays and explains their desirability for many edible fungi.

Marteka, V. 1980. Growing your own mushrooms. *Mushrooms: Wild and edible*, 245–47. New York: W. W. Norton. Covers in a general way growing mushrooms from spores, from spawn outdoors, indoors, and from kits.

Quimio, T. H., S. T. Chang, and D. J. Royse. 1990. *Technical guidelines for mushroom growing in the tropics*. FAO Plant Prod. and Prot. Paper 106, FAO, Rome. Decent coverage of a number of mushrooms that can also be produced in California.

Rinaldi A., and V. Tyndalo. 1974. Gastronomic classification of mushrooms, and How to preserve mushrooms. *Mushrooms and other fungi*, 253–54 and 255–56. London: Hamlyn Pub. The first article classifies mushrooms by their taste and uses, and the second includes a list of mushrooms that can be preserved in oil and vinegar.

Sandoval, R. 1985. Mushrooms sprout in Sierra foothills. *Ag Alert* 12(38):10–11.

Sims, William. 1979. *Growing mushrooms*. University of California Cooperative Extension, Leaflet 2640.

Stamets, P., and J. S. Chilton. 1983. *The mushroom cultivator: A practical guide to growing mushrooms at home*. Olympia, Wash.: Agarikon Press. With the exception of truffles, gives very good coverage of the edible fungi.

Wells, M. 1989. The cheapo incubator . . . one you can build. *Mushroom* 7(2):25–26. Small-scale incubators for $10 to $40.

Wells, M. 1989. For culture work you need a sterile environment. *Mushroom* 7(3):20–22. Includes agar medium formulas, equipment needed, and step-by-step formulation.

Wells, M. 1990. It's time to spawn. *Mushroom* 8(1):23–25.

Wells, M. 1990. If you coax along some species, they'll behave like perennials. *Mushroom* 8(2):22–24.

Wells, M. 1990. Filters on the cheap. *Mushroom* 8(3):20. Using coffee and milk filters rather than the commercial spawn jar lid filter.

Wells, M. 1990. Can you sterilize with a microwave? *Mushroom* 8(3):20–21. A possibility of replacing the pressure cooker.

Wells, M. 1991. Three+ recipes with potatoes. *Mushroom* 9(2):37. Four recipes (three serious) for making potato dextrose agar.

Cuitlacoche, Mexican Truffle, Corn Smut (*Ustilago maydis*)
When harvested at the proper time, the smut is a delicacy now popular in some of the higher-class restaurants in the eastern United States and in Mexico. Probably should not be advertised as "corn smut."

Kennedy, D. 1978. *Recipes from the regional cooks of Mexico*, 141. New York: Harper & Row. Has one recipe using zucchini and smut.

Pataky, J. K. 1991. Production of cuitacoche (*Ustilago maydis*) on sweet corn. *HortSci.* 26:1374–77. The results of artifical induction of smut were pretty abysmal, but proceedings are described. It also lists hybrid sweet corns and their percentage of natural infection over three years.

Thakur, R. P., et al. 1989. Smut gall development in adult corn plants inoculated with *Ustilago maydis*. *Plant Dis. Rep.* 73:921–25. Considerably more successful with inoculation.

Wilcoxson, R. D. 1975. The relationship between corn plant population and smut infections. *Plant Dis. Rep.* 59(8):678–80. The lower the plant population, the higher the incident of smut infection.

Enoke, Winter Mushroom, Velvet Stem (*Flammulina velutipes*)
A good-tasting, wood-decaying fungus that needs autoclaved substrates to produce an adequate crop.

Chang, S. T., and P. G. Miles. 1989. *Flammulina* and *Pholiota: Low-temperature cultivated mushrooms.* 255–58.*Edible mushrooms and their cultivation*, 261–62. Boca Raton, Fla.: CRC. Fairly detailed with some pointers not in *The mushroom cultivator*.

Stamets, P., and J. S. Chilton. 1983.*The mushroom cultivator: A practical guide to growing mushrooms at home*, 172–75. Olympia, Wash.: Agarikon Press. With references to other sections. Better presentation of the growing parameters.

Garden Giant, Winecap Stropharia (*Stropharia rugosa-annulata*)
Chilton, J. S. 1987. The garden giant can be cultivated. *Mushroom* 5(1):17–19.

Wells, M. 1990. Earthless winecap stropharia. *Mushroom,* 8(3):20. It may need a spring climate similar to Olympia, Washington.

Ink Cap (*Coprinus comatus*)
Very common spring mushroom in horse manure. Good flavor but breaks down quickly.

Lelley, L. 1983. Investigations on the culture of the ink cap. *Mushroom J.* 129[130]:364–70.

Nameko, Viscid mushroom (*Pholiota nameko*)
This good-flavored and pleasingly colored, low-temperature mushroom can be fairly easily grown on semipasteurized, high-lignon substrates like straw and sawdust.

Chang, S. T., and P. G. Miles. 1989. *Flammulina* and *Pholiota*: Low-temperature cultivated mushrooms. *Edible mushrooms and their cultivation*, 258–61, 263. Boca Raton, Fla.: CRC.

Krishna, A., and B. K. Sharma. 1987. Domestication of Nameko mushroom in India. *Mushroom Sci.* 12(2):479–83.

Oyster (*Pleurotus* spp.)

One species of this fan-shaped mushroom, frequently found growing on trees or deadwood, can be cultivated on any number of high-lignin agricultural wastes like corn cobs, straw, sawdust, sunflower stalks. Becoming increasingly popular in the restaurant trade and at farmers' markets.

Bano, Z., S. Rajarathnum, and N. Nagaraja. 1978. Some aspects of the cultivation of *Pleurotus flabellatis* in India. Covers "substrate spawn" from hot-water-treated paddy straw and supplementing spawn beds with cotton seed and horse gram (*Dolichos biflorus*) powder.

Brooke-Webster, D., and A. A. Cairns. 1986. The use of polyethylene film to control the fructification of *Pleurotus* spp. grown on horizontal trays. In P. G. Wuest, et al. (eds.), *Cultivating Edible Fungi*, 433–36. Amsterdam: Elsevier Sci. Pub.

Chang, S. T., O. W. Lau, and K. Y. Cho. 1981. The cultivation and nutritional value of *Pleurotus sajer-caju*. *Eur. J. Appl. Microbiol. Biotech*. 12:58–62. Uses both cotton waste and straw and fairly simple techniques.

Ganeshan, G., R. P. Teware, and B. S. Bhargava. 1987. Influence of residual vegetable crop biomass on yield and mineral content of *Pleurotus sajor-caju*. Discusses the use of several vegetable crop residues as substrates. The implications are that other residues would also work. Should be useful to vegetable growers.

Kandaswamy, T. K., and K. Ramasamy. 1976. Effects of organic substrates with different C:N ratio on the yield of *Pleurotus sajor-cajo*. In *Indian Mush. Sci*. (1):423–27. Presents possible substrate materials including vegetable crops wastes.

Lanze, G. 1986. The cultivation of the oyster mushroom in bags—with or without holes. *Mushroom Information* 3(4):36–40.

Miller, R. A. 1988. The cultivation and economics of oyster mushrooms. *Herb Mark. Rep*. 4(12):5–6.

Pettipher, G. L. 1987. Cultivation of the oyster mushroom on lignocellulosic waste. *J. Sci. Food Agric*. 41:259–65. Uses simple methods and equipment and cocoa hull waste but says straw would work better.

San Antonio, J. P., and P. K. Hanners. 1984. Using basidioshores of the oyster mushroom to prepare grain spawn. *HortSci*. 19:684–86. A fairly simple "home remedy" for producing grain spawn designed for the small grower.

Sharma, V. P., and C. L. Jandaik. 1991. Effects of supplementation on yield and size of *Pleurotus sajor-caju*. *Mushroom Info*. 8(6):12–15, 18–20. Uses common waste products to increase yields.

Verma, R. M., G. B. Singh, and S. M. Singh. 1991. Spawn cycling: A new approach to cultivation of oyster mushrooms. *Mushroom Info*. 8(6):6–10. A simple method of "cultivating" own spawn, thus eliminating necessity of purchasing new spawn for each production cycle.

Wells, M. 1990. Kicking and screaming, we're dragged back to *Pleurotus*. *Mushroom* 8 (3):19–20. Yet another slightly more labor-intensive way to grow *Pleurotus* with higher yields.

Wells, M. 1989. *Pleurotus* grow fast, so your span of attention isn't tried. *Mushroom*, 7(1):19–21. The breezier presentation makes it seem a lot easier.

Zadrazil, F. 1980. Influence of ammonium nitrate and organic supplements on the yield of *Pleurotus sajor-caju*. *Eur. J. Appl. Microbiol. Biotech*. 9:31–35.

Zadrazil, F. 1985. Basis for *Pleurotus* cultivation. *Mushroom Info*. 2(7/8):26–33 and 2(10):10–19.

Shiitake (*Lentinus edodes*)

Traditionally grown in Japan on inoculated hardwood logs, amended sawdust, straw, and other materials are increasingly used as substrates. Cultural practices range from low-volume "sideline" production to "high-tech" high-volume primary business. Growing demand in the specialty food markets.

Akiyamo, H., et al. 1974. The new cultivation of shiitake in a short period. In K. Mori (ed.), *Mushroom Sci*. 9(1):423–33. Whereas traditional methods require waiting one and one-half to two years after inoculating the logs, the methods described here require only six months and allow cultivation in more climatic areas.

Chang, S. T., and P. G. Miles. 1989. Lentinus: A mushrooming mushroom. *Edible mushrooms and their cultivation*, 194–223. Boca Raton, Fla.: CRC. Quite detailed, including usable trees and formulas for synthetic substrates.

Cotter, H., and T. Flynn. 1986. Home shiitake: An easy, inexpensive cultivation method. *Mushroom* 4(4):35–36.

Cotter, H., and T. Flynn. 1987. Evaluation of shiitake log inoculation techniques. In K. Grabbe and O. Hilber (eds.), *Mushroom Sci.* 12(2):293–301. Evaluates drilling vs. chain saw kerf inoculation and standing tree inoculation in W. Virginia.

Delpech, P., and J. M. Oliver. 1990. The perfumed mushroom (or shiitake): A French method of cultivating this species. *Mushroom Info.* 7(9):4–5 and 7(10):4–6. This method uses straw as the substrate.

Harris, Bob. 1986. *Growing shiitake commercially: A practical manual for the production of Japanese forest mushrooms*. Madison, Wis.: Science Tech. Pubs.

Ho, M. S. 1987. A new technology: "Plastic bag cultivation method" for growing shiitake mushrooms. *Mushroom Sci.* 12(2):303–307. Gives detailed steps for a process that can use the sawdust of a fairly wide variety of trees.

Juliano, J. 1989. Shiitake mushroom spawn: Production, quality, storage, strain selection and development. In *Shiitake mushrooms: A national symposium and trade show*, 29–46. St. Paul: University of Minnesota. Very detailed and well-illustrated presentation of this important aspect.

LaBorde, J. 1991. A glance at shiitake production for 1991. *Mushroom Info.* 8(11):5–15. Fairly general but with some good methods briefly discussed.

Leatham, G. F. 1982. Cultivation of shiitake, the Japanese forest mushroom, on logs. *Forest Prod. J.* 32(8):29–35.

Pettipher, G. L. 1988. Cultivation of the shiitake mushroom on lignocellulosic waste. *J. Sci. Food Agric.* 42:195–98. Rather scientific presentation but reference below adds some needed detail.

Pettipher, G. L. 1988. Cultivation of the oyster and shiitake mushrooms on lignocellulosic wastes. *Mushroom J.* 183:491, 493.

Royse, D. J. 1985. Effect of spawn run time and substrate nutrition on yield and size of shiitake mushrooms. *Mycologia* 77:756–62. Nutritional supplements in sawdust substrate and longer incubation periods lead to higher production rates and increased mushroom size. In addition, the spawn run is easier to control. But results are also dependent on genotypes of parental shiitake.

Royse, D. J. 1986. Effects of genotype, spawn run time and substrate formulation on biological efficiency of shiitake. *Appl. Environ. Microbiol.* 52:1425–27. Need to use in conjunction with the Pettipher reference, "Cultivation of the oyster and shiitake mushrooms . . . ," and again, genotypes produce different results.

Royse, D. J., B. D. Bahler, and C. C. Bahler. 1990. Enhanced yield of shiitake by saccharide amendment of the synthetic substrate. *Appl. Environ. Microbiol.* 56:479–82. Amending synthetic sawdust-wheat bran-millet substrate with selected saccharides led to higher mushroom yields. Recommends using 1.2 percent sucrose because it is more readily and cheaply available.

San Antonio, J. P. 1981. Cultivation of the shiitake mushrooms. *HortSci.* 16:151–56. Presents a very "crude" open-air, log-inoculated procedure. The important variables are temperature and precipitation, of which the latter could be limiting in California. Recommends this as a small farm supplemental crop.

Stamets, P., and J. S. Chilton. 1983. *The mushroom cultivator*, 176–79 (with references to other sections of the book). Olympia, Wash.: Agarikon Press. Somewhat detailed presentation of the growth parameters.

Wilkes, G. 1985. Shiitake: A new forest product. *Am. For.* 91(10):48–49. Briefly covers outdoor cultivation but has some practical suggestions.

Wuest, P. J. 1989. Shiitake growing in sawdust. In *Shittake mushrooms: A national symposium and trade show*, 47–55. St. Paul: University of Minnesota.

Straw (*Volvariella volvacea*)
The traditional mushroom of Southeast Asia is fairly easy to cultivate on a much wider variety of agricultural waste substrates.

Chang, S. T., and P. G. Miles. 1989. *Volvariella*: A high-temperature cultivated mushroom. *Edible mushrooms and their cultivation*, 226–53. Boca Raton, Fla.: CRC. Quite detailed but emphasizes substrates (such as cotton wastes) not readily available in most of California.

Stamets, P., and J. S. Chilton. 1983. *The mushroom cultivator*, 214–16 (with references to other sections of the book). Olympia, Wash.: Agarikon Press. Somewhat detailed presentation of growing parameters with emphasis on more easily obtained substrate materials.

Sukara, E. 1985. The cultivation of the paddy straw mushroom. *Bull. Brit. Mycol. Soc.* 19(2):129–32. Presents a fairly simple outdoor method used in Indonesia.

Truffle (*Tuber* spp.)
Highly regarded, and hence highly priced, especially in France and Italy, it is now possible to "cultivate" this previously wild underground fungus on inoculated trees.

Korn, L. 1987. Raising American truffles. *Org. Gard.* 34(10):24, 26–27. Gives the basics as well as sources of inoculated seedling trees and the address of the N. American Truffling Society.

Singer, R., and B. Harris. 1987. Cultivation of the perigord truffle, and Appendix IV: Truffles. *Mushrooms and truffles*, 196–202 and 322–25. Koenigstein, W. Germany: Koeltz Sci. Books. The appendix is much more thorough.

White Jelly Fungus, "Silver Ear" Mushroom (*Tremella fuciformis*)
Chang, S. T., and P. G. Miles. 1989. *Tremella*-increased production by a mixed culture technique. *Edible mushrooms and their cultivation*, 277–91. Boca Raton, Fla.: CRC. Presents several cultivation techniques for this Chinese medicinal and culinary mushroom. Obtaining spawn could be a problem.

Woodear (*Auricularia* spp.)
A black gelatinous "mushroom" grown on logs or sawdust, used by the Chinese and Southeast Asians dried, pickled, fried, steamed, or boiled. Not too popular with Westerners.

Lou, L. H. 1981. Woodear cultivation in plastic sheds around Peking. *Mushroom Sci.* 11(1):711–16.

Quimio, T. H. 1981. Philippine *Auricularias*: Taxonomy, nutrition and cultivation. *Mushroom Sci.* 11 (2):685–96.

NUTS

Filbert (*Corylus avellana*)
Deciduous trees 15 to 25 feet tall, bearing round to oblong nuts.

Bassil, N. V., et al. 1991. Propagation of hazelnut stem cuttings using *Agrobacterium* rhizogenes. *HortSci.* 26(8):1058–60. Specific strains or mixed strains of the bacterium cause better rooting and bud retention.

Childers, N. F. 1983. Filberts. *Modern fruit science*, 303–6. Gainesville, Fla.: Horticultural Pub.

Lagerstedt, H. 1979. Filberts. In R. A. Jaynes (ed.), *Nut tree culture in North America*, 128–47. Hamden, Conn.: North America Nut Growers Assoc.

Macadamia nut
Evergreen tree, 25 to 30 feet tall by 15 to 20 feet wide, that starts producing nuts in three to five years.

Thomson, P. H. 1970. Home processing of macadamia nuts. *CFRG Yearbook* 2:64–66.

Thomson, P. H. 1979. Macadamia. In R. A. Jaynes (ed.), *Nut tree culture in North America*, 188–202. Hamden, Conn.: North America Nut Growers Assoc.

Thomson, P. H. 1980. The macadamia in California. *CRFG Yearbook* 12:46–115.

Pecan (*Carya illinoinensis, C. pecan*)
Tall (up to 70 feet) and wide deciduous tree, producing (ideally) 50 pounds of nuts per tree each year from age 10 to 15 years.

Madden, G. 1979. Pecans. In R. A. Jaynes (ed.), *Nut tree culture in North America*, 13–34. Hamden, Conn.: North America Nut Growers Assoc.

Childers, N. F. 1983. *Carya* species. *Modern fruit science*, 283–94. Gainesville, Fla.: Horticultural Pub.

Herrera, E. 1991. 1991. Boosting pecan size: Properly timed watering can improve nut size. *Nut Grower* 11(1):20, 28. The whys and hows of irrigation.

House, J. 1991. Saving pecan trees: Transplanting means shorter waits for production. *Nut Grower* 11(1):6–7, 10.

Sibbet, S., et al. 1989. Evaluating pecans for California. *Nut Grower* 8(4):24–26, 29. Details on why pecans show good promise.

Pine nuts (*Pinus* spp.)
Millikan, D. F. 1979. Pines. In R. A. Jaynes (ed.), *Nut tree culture in North America*, 183–86. Hamden, Conn.: North America Nut Growers Assoc.

ONIONS

Boiler, pickler (*Allium cepa* spp.)
Voss, R. E. 1979. Onion production in California. University of California Cooperative Extension, Priced Publication 4097, pp. 38–40. Very complete for all aspects of California onion production.

Japanese bunching (Welsh onion) (*Allium fistulosum*)
Harrington, G. 1978. Multiplier onions. *Grow your own Chinese vegetables*, 165–69. Pownal, Vt.: Garden Way Publishing. Detailed instructions for perennial green onions.

Jones H. A., and L. K. Mann. 1963. *Onions and their allies: Botany, cultivation, and utilization*, 239–41. New York: Interscience Publishers. Deals mainly with very labor-intensive methods of producing blanched stems and also discusses several varieties.

Rakkyo (*A. chinese*)
Densely clumped, chive-like in growth, with crisp, textured bulbs with strong onion-like but distinctive odor. Used both fresh and pickled.

Jones H. A., and L. K. Mann. 1963. *Onions and their allies: Botany, cultivation, and utilization*, 241–44. New York: Interscience Publishers.

Mann, L. K., and W. T. Stearn. 1960. Rakkyo or ch'iao t'ou: A little known vegetable crop. *Econ. Bot.* 14:69–83.

ORACH, GARDEN (*Atriplex hortensis*)
Darlsson, R., and C. W. Hallquist. 1981. Atriplex hortensis: Revival of a spinach plant. Acta Agric. Scand. 31:229–31.

Halpin, A. (ed.). 1978. *Unusual vegetables: Something new for this year's garden*, 279–81. Emmaus, Penn.: Rodale Press.

PALM, JELLY (*Butia capitata*)
Slow-growing, 10-to 20-foot hardy palm with large clusters of 1-inch-diameter, orange, globular fruitlets that are very fragrant and eaten out of hand or as a jam, having a somewhat pineapple-like flavor.

Givan, R. 1988. *Butia capitata*. Pomona 21(1):16–17.

Riley, J. M. 1987. The jelly palm. *Fruit Gard.* 19(1):10.

PAPAYA (*Carica papaya*), Mountain papaya
IC&RF Soc. 1989. What they say about papaya. *IC&RF Soc. Newsl.* 29:8–9.

Lievins, B., and J. Neitzel. 1979. The papaya. *CRFG Newsletter* 11(2):10–14.

Muthukrishnan, C. R., and I. Irulappan. 1990. Papaya. In T. K. Bose and S. K. Mitra (eds.), *Fruits: Tropical and subtropical*, 304–35. Calcutta: Naya Prokash. The others are more useful overall, but this has good information among the wealth of details.

Okamoto, E. K., and F. Cooper. 1970. Papaya section. *CRFG Yearbook* 2:7–16.

Riley, J. M. 1976. The papaya and its relatives. *CFRG Yearbook* 8:112–18.

PARSNIP (*Pastinaca sativa*)
Brooks, W. M. 1972. Vegetables with edible roots. *Plants Gard.* 28(2):57–59.

PASSIFLORA spp. (Passion Fruit)
Perennial 20- to 30-foot vines with small egg-shaped, juicy fruits. See individual listings below.

Banana passion fruit (*P. mollissima*)
Fruits to 3 inches long, banana shaped; not good for juice.

Morton, J. F. 1987. *Fruits of warm climates*, 332–33. Winterville, N.C.: J. F. Morton.

Maypop (*P. incarta*)
Fisch, M. B., and F. A. Kuhne. 1975. Passion fruit and granadilla section. *CRFG Yearbook* 7:13–71.

Menzel, C. M., et al. 1989. Passionfruit in Queensland. 3. Orchard management. *Queensl. Agric. J.* 115:155–64. Very practical, but remember the seasons are reversed in Australia.

Morton, J. F. 1987. *Fruits of warm climates*, 320–28. Winterville, N.C.: J. F. Morton.

Passion fruit, purple granadilla (*P. edulis*)
Yellow giant granadilla, Yellow passion fruit (*P. edulis forma flavicarpa*)
Requires more warmth for fruiting.

PAW PAW (*Asimina triloba*)
Extremely hardy, 25-foot deciduous tree with 3-to 5-inch-long, cylindrical, brown-black fruit with very aromatic banana-flavored, creamy-yellow, sweet, melting pulp.

Paw Paw section. 1974. *CRFG Yearbook* 6:10–206. (Condensed version of above. 1982. *CRFG Yearbook* 14:5–46.)

Rare fruit chart. 1987. *IC&RF Soc. Newsl.* 23. Gives nursery sources.

PEARS

Asian (*Pyrus pyrifolia, P. ussuriensis*)
Griggs, William H., and Ben T. Iwakiri. 1977. *Asian pear varieties in California*. University of California Cooperative Extension, Priced Publication 4068.

Beutel, J. A. 1988. *Asian pears*. Family Farm Series 1/88, Small Farm Center, University of California, Davis, p. 6.

European (*Pyrus communis*)
Kaiser, P. 1988. Two very good old pears: Hardy and superfine. *Pomona* 22(1):51–52.

PEAS, EDIBLE POD (*Pisum sativum* var. *macrocarpor*)
Harrington, G. 1985. Snow peas. *Grow your own Chinese vegetables*, 65–71. Pownal, Vt.: Garden Way Publishing. Gives a few more helpful hints.

Valenzuela, L. 1983. *Edible pod pea production in California*. University of California Cooperative Extension, Leaflet 21328.

PECANS—See *Nuts: Pecans*

PEPINO (*Solanum muricatum*)
Perennial evergreen shrub often treated as an annual, with fruit that is 1 to 2 inches long, eggplant-shaped and green- to cream-colored with purple stripes. Has light-green to whitish flesh and a tomato-melon flavor.

Baker, H. 1989. About the pepino. *IC&RF Soc. Newsl.* 28:15.

Sweet, C. 1986. Pepino: Can it grow here? *Calif. Grower: Avocados, Citrus, Subtropicals* 10(7):19, 34.

PEPPERS (*Capiscum annum, C. frutescens*)
Hall, H., S. Wada, and R. Voss. 1975. *Vegetable gardening: Growing peppers*. University of California Cooperative Extension, Leaflet 2773.

Sims, W., and P. G. Smith. 1971. *Growing peppers in California*. University of California Cooperative Extension, Leaflet 2676.

Whealy, K. 1988. *The garden seed inventory*. Decorah, Ia.: Seed Saver Publications. For unusual nonhybrid varieties available in 1987, especially for those with unusual colors. For hybrids and post-1987 nonhybrid varieties, browse general and "specialty" seed catalogs.

Paprika (*Capsicum annum*)
Miller, R. 1991. The cultivation and harvest of paprikas. *Herb Mark. Rep.* 7(3):1–6, (especially pp. 4–6).

PERSIMMON

American (*Diospyrus virginiana*)
Claypool, J. 1988. Improving the American persimmon: A progress report. *Pomona* 21(2):7–11.

Gleason, J. K. 1991. The beginning: A seed. *Pomona* 24(1):11.

IC&RF Soc. 1988. Persimmons. *IC&RF Soc. Newsl.* 27:8–9. Brief information on American and several varieties of Japanese persimmons.

Morton, J. F. 1987. *Fruits of warm climates*, 411–16. Winterville, N.C.: J. F. Morton.

General
LaRue, J., K. Opitz, and J. Beutel. 1982. *Growing persimmons in California*. Berkeley: University of California Division of Agricultural Sciences, Leaflet 21277.

Japanese (*Diospyros kaki*)
Long-lived, 15- to 60-foot tree with variously shaped and colored fruit with soft, sweet flesh when ripe; some dark-fleshed types crisp and sweet before fully ripe. Can be eaten out of hand, added to salads, ice cream, and other dishes, made into jam or marmalade, frozen, dried, or stewed.

George, A. P., and R. J. Nissen. 1990. Persimmon. In T. K. Bose and S. K. Mitra (eds.), *Fruits: Tropical and subtropical*, 469–89. Calcutta: Naya Prokash. Useful added detail.

PHALSA (*Grewia subinaequalis* and *G. asiatica*)
Dwarf shrubs with unevenly ripening, ½- to ⅝-inch dark-purple fruits with soft, fibrous, purple-red flesh. It has a somewhat grape-like flavor and can be eaten fresh or made into a syrup or juice. As long as soil is not waterlogged, will grow in areas where other fruits won't, including moderately alkaline soils.

Morton, J. F. 1987. Fruits of warm climates, 276–77. Winterville, N.C.: J. F. Morton.

Shankar, G. 1990. Phalsa. In T. K. Bose and S. K. Mitra (eds.), *Fruits: Tropical and subtropical*, 774–80. Calcutta: Naya Prokash. Quite detailed.

PINE NUTS—See *Nuts: Pine nuts*

PITANGA—See Surinam cherry

PLUM, NATAL—See Carissa

POHA BERRY (*Physalis peruviana*)
Herbaceous perennial, 2 to 6 feet tall, treated as an annual with husk-covered, ½- to ¾-inch, smooth, orange-yellow berry with juicy, sweet pulp. See recipes in C. Dremann.

Dremann, C. 1985. *Ground cherries, husk tomatoes, and tomatillos*, 1–2, 7–10. Redwood City, Calif.: Redwood City Seed Company.

Englehart, O. H. 1986. The poha berry or husk berry. *CRFG Newsletter* 18(2):29.

Sweet, C. 1986. Cape gooseberry: It can grow here, but does anyone care? *Calif. Grower* 10(8):26–28. Thorough growing instructions.

POMEGRANATE
Tall (20 to 30 feet), long-lived, somewhat spiny small trees with 2½- to 5-inch round, tough-skinned, yellow fruit packed with transparent sacs filled with tart, flavorful, fleshy, juicy, red pulp and seeds that are eaten out of hand or juiced.

LaRue, J. A. 1977. *Growing pomegranates in California*. University of California Cooperative Extension, Leaflet 2459.

Morton, J. F. 1987. *Fruits of warm climates*, 352–55. Winterville, Fla.: J. F. Morton.

Patil, A. V., and A. R. Karale. 1985. In T. K. Bose, *Fruits of India: Tropical and sub-tropical*, 537–48. Calcutta: Naya Prokash.

Patil, A. V., and A. R. Karale. 1990. Pomegranate. In T. K. Bose and S. K. Mitra (eds.), *Fruits: Tropical and subtropical*, 616–37. Calcutta: Naya Prokash. More detailed than the 1985 edition.

Daru
A very hardy, erect, deciduous, spreading 15- to 24-foot tree producing up to 70 pounds of fruit per tree. The fruit seeds are dried to produce a widely used, expensive sour spice.

Parmar, C. 1988. Daru: A wild pomegranate from the Himalayas. *Pomona* 21(3):33–35.

POTATO (*Solanum tuberosum*)
Drowns, G. 1990. Special spuds. *Fine Gard.* 13:66–71. Article complete with insect control, special varieties, sources.

Pollock, F. 1991. Tiny (buried) treasures. *Org. Gard.* 38(9):26–32. Good pointers for growing "baby" potatoes.

Staff. 1989. Potato surprises: New colors, shapes, flavors. *Sunset* 182(1):64–65.

Staff. 1989. Where to buy the less usual potatoes. *Sunset* 182(1):132.

Whealy, K. (ed.), *The garden seed inventory.* Decorah, Ia.: Seed Saver Publications. For unusual nonhybrid varieties available in 1987. For hybrids and post-1987 nonhybrid varieties, browse general and "specialty" seed catalogs.

PUMMELO (*Citrus maxima*)

Tall (16 to 20 feet) citrus with yellowish, round- to pear-shaped, 4- to 12-inch-wide fruit with very juicy to fairly dry, mild, and sweet- to subacid-flavored pulp that is usually eaten out of hand, but also in salads and desserts or as a preserve. The peel can be candied. Grown at experiment stations in Indio and Riverside, California.

Morton, J. F. 1987. Pummelo. *Fruits of warm climates*, 147–51. Winterville, N.C.: J. F. Morton.

PURSLANE (*Portulaca oleracea* subsp. *sativa*)

Called *pourpier* in France and *verdolaga* in Mexico, it has a succulent texture and mildly ascerbic flavor. It is used both raw and cooked.

Gorman, M. 1988. Purslane goes American. *Org. Gard.* 35 (6):68–69.

QUINCE (*Cydonia oblonga*)

Deciduous, slow-growing trees to 15 feet tall, producing from mid-October to early November with strongly aromatic, well-ripened fruit used for jellies, jams, and preserves.

Lisle, H. C. 1991. Propagating quince from seed. *Pomona* 24(1):35–56.

Smith, A. H. 1977. The quince. *CRFG Yearbook* 9:50–69.

QUINOA (*Chenopodium quinoa*)

Can't take high temperatures at blooming or seed set and needs a lot of light and short days to bloom. Best in high mountains; some strains adaptable to sea level.

Abundant Life Seed Catalog, pp. 19–20. See Seed Sources, p. 33. Brief Cultivation instructions and available strains.

Quinoa Corporation, P.O. Box 7114, Boulder, CO 80306. Price list of bulk grain and further information.

Wood, R. 1988. *Quinoa: The supergrain.* Briarcliff Manor, N.Y.: Japan Pubns. USA. Chiefly a recipe book, but has a section on cultivation.

RADICCHIO—See *Chicory: Radicchio*

RADISH (*Raphanus sativus*)

Whealy, K. 1988. *The garden seed inventory.* Decorah, Ia.: Seed Saver Publications. For unusual nonhybrid varieties available in 1987. For hybrids and post-1987 nonhybrid varieties, browse general and "specialty" seed catalogs.

Chinese (*Raphanus sativus* var. *longipinnatus*)

Harrington, G. 1985. *Grow your own Chinese vegetables*, 117–22. Pownal, Vt.: Garden Way Publishing.

RAISIN TREE (*Hovenia dulcis*)

An irregularly fruiting tree to 70 feet wide by 80 feet tall, whose enlarged flower stalks, after frost, are filled with yellowish, pear-flavored pulp.

Koller, G. L., and J. H. Alexander III. 1979. The raisin tree: Its use, hardiness and size. *Arnoldia* 39(1):6–15.

Quesada, G. C. 1990. Germination requirements for *Hovenia dulcis* seeds. *Fruit Gard.* 22(4):45. Discusses the scarification and stratification periods needed to germinate the seed.

Riley, J. M. 1970. The Japanese raisin tree. *CRFG Newsletter* 2(2):1–2, 5. Provides information in relation to California.

RASPBERRIES—See *Berries: Raspberries*

RHUBARB (*Rheum rhaponticum*)

The first three of the following articles together present the basics for growing and using this perennial.

Carlsen, K. L. 1970. Is it time to move the rhubarb? *Org. Gard.* 17(4):100–101.

Carlsen, K. L. 1982. Don't move the rhubarb! *Org. Gard.* 29(3):29–30.

Hertzog, L. H. 1978. Rhubarb pays off. *Org. Gard.* 25(4):68–69.

Johnson, W. B. 1972. Asparagus and rhubarb. *Plants Gard.* 28(2):62–63.

ROCKET—See *Arugula*

ROSE APPLE (*Syzygium jambos*)
Dense, 30-foot evergreen tree with hollow, egg-shaped, yellow or pinkish-white fruit with thin, crisp, rose-flavored flesh. Can be eaten out of hand, candied, or made into jellies.

Fisch, B. E. 1976. The rose apple. *CRFG Yearbook* 8:100–111.

ROSELLE (*Hibiscus sabdariffa*)
Annual, erect, subshrub to 8 feet tall with red, 1¼- to 2¼-inch, fleshy, juicy, crisp calyxes with a cranberry taste used chopped in salads or made into sauces or juices.

Morton, J. F. 1987. *Fruits of warm climates*, 281–86. Winterville, N.C.: J. F. Morton.

Riley, J. M. 1973. The roselle. *CRFG Yearbook* 5:102–4.

RUTABAGA (*Brassica napus* var. *napobrassica*)
Brooks, W. M. 1972. Plants with edible roots. *Plants Gard.* 28(2):58–59.

SALSIFY (*Tragopogon porrifolius*)
Brooks, W. M. 1972. Vegetables with edible roots. *Plants Gard.* 28(2):57–59.

SAPODILLA (*Manilkara zapota*)
Evergreen tree to 50 feet tall, with 5-inch-diameter, oval fruit with thin, rough, gray to brown skin and yellow-brown, aromatic, juicy flesh that is eaten out of the shell or made into sauce, syrup, jam, pies, cakes, as well as dried. With protection, a tree is currently said to be producing in the Sacramento valley. Mature trees can stand several hours of 26° to 28°F temperatures.

Campbell, C. W., et al. "Tikal": An early-maturing sapodilla cultivar. *Trop. Fruit News* 22 (3):29–30.

Joyner, G. 1988. Fruit of the month: Sapodilla. *Trop. Fruit News* 22(7):66. Has a few details not covered by J. F. Morton.

Morton, J. F. 1987. *Fruits of warm climates*, 393–98. Winterville, N.C.: J. F. Morton.

Rare fruit chart. 1987. *IC&RF Soc. Newsl.* 23:12. Gives sources of plants.

SAPOTE

Black (*Diospyros dignya*, also *D. ebenaster*)
Evergreen tree to 25 feet tall with round, 5-inch diameter, greenish-brown fruit with chocolate-pudding taste. Pulp is served as a dessert, on ice cream, or used in pie filling.

Fisch, B. E. 1975. Black sapote. *CRFG Yearbook* 7:91–97.

Morton, J. F. 1987. *Fruits of warm climates*, 416–18. Winterville, N.C.: J. F. Morton.

Green (*Calocarpum viride*)
Slow-growing tree to 40 feet tall, with very thin-skinned, yellowish-green to dull brownish-orange fruit having soft, bright orange-red, thick, custard-consistency flesh with a very sweet, squash-like flavor.

Thomson, P. H. 1973. The green sapote. *CRFG Yearbook* 5:41–48.

Mamey (*Pouteria sapota*)
Open, tall (to 90 feet) tree with egg-shaped, russet-brown, rough-skinned fruit up to 8 inches long, with slightly sticky, sweetish flesh. Can be eaten out of hand, in sherbet, as a tart, or as a marmalade. The kernels are also edible with minimum processing.

Fisch, M. B. 1973. The mamey or red sapote. *CRFG Yearbook* 5:24–34.

IC&RF Soc. 1987. 10-page illustrated rare fruit chart. *IC&RF Soc. Newsl.* 23:13. Gives sources of plants.

Morton, J. F. 1987. *Fruits of warm climates*, 398–401. Winterville, N.C.: J. F. Morton.

Thomson, P. H. 1973. Notes on the mamey sapote. *CRFG Yearbook* 5:39–40.

Whitman, W. F. 1973. The mamey sapote in Florida. *CRFG Yearbook* 5:35–38.

White (*Casimiroa edulis*)

Fast-growing evergreen tree to 60 feet tall, with yellow, tomato- to orange-sized fruit with thin papery skin and very sweet and juicy soft white pulp that is eaten out of hand, added to salads, fruit cups, ice cream, milk shakes, and made into a marmalade.

Chambers, R. R. 1984. White sapote varieties: Progress report. *CRFG Yearbook* 16:56–64.

Englehart, O. H. 1977. Reviewing *Casomiroa edulis*. *CRFG Yearbook* 9:35–36. Promising varieties for Southern California.

Morton, J. F. *Fruits of warm climates*, 191–96. Winterville, N.C.: J. F. Morton.

Thomson, P. H. 1973. The white sapote. *CRFG Yearbook* 5:6–20.

Yellow (*Casimiroa tetrameria*)

Tall (15 to 20 feet) trees that withstand neglect and drought better, have greenish-yellow to dull, deep yellow, tougher-skinned fruit with spicy, delicate-flavored, golden flesh.

Ramsey, C. W. 1973. The yellow sapote. *CRFG Yearbook* 5:21–23.

SEAKALE (*Crambe maritina*)

During the first year, but preferably the second, the leaf stalks are blanched to obtain a winter vegetable that, when boiled, has a slightly bitter, hazelnut flavor.

Peron, J. Y. 1988. Seakale: A new vegetable produced as etiolated sprouts. In J. Janick and J. E. Simon (eds.), *Advances in new crops*, 419–22. Portland, Ore.: Timber Press.

Vilmorin-Andeieux, Mm. 1976. *The vegetable garden*, 516–22. Palo Alto, Calif.: Jeavons-Leler. This reprint of an 1885 work gives detailed directions for growing "the old-fashioned way."

SHALLOTS—See *Herbs: Allium – Shallots*

SPROUTS

Larimore, B. B. 1975. *Sprouting for all seasons: How and what to sprout, including delicious, easy to prepare recipes*. Bountiful, Ut.: Horizon Publishers.

STAR FRUIT—See *Carambola*

SUBTROPICAL FRUIT—See *Fruits: Temperate*

SURINAM CHERRY or PITANGA (*Eugenia uniflora*)

Fast-growing evergreen, to 15 feet tall, with deep-red, small, tomato-sized fruit with soft, juicy, sweetly tart flavor that is eaten out of hand or in pies and jellies.

Jerris, W. V. 1989. The pitanga: Home use and potential commercial applications. *Fruit Gard.* 21 (11):5–9.

Morton, J. F. 1987. *Fruits of warm climates*, 386–86. Winterville, N.C.: J. F. Morton.

Westree, N. E., and J. Riley. Pitanga section. *CRFG Yearbook* 3:9–25.

TAMARILLO (*Physalis ixocarpa*) or Tree Tomato

Fast-growing evergreen shrub to 13 feet tall with 2-inch-long, egg-shaped fruits that can be used in cooking like a tomato, eaten raw with sugar, added to pies, jams, or chutney.

Fisch, M. B. 1974. The tree tomato. *CRFG Yearbook* 6:268–90.

Morton, J. F. 1987. Tree tomato. *Fruits of warm climates*, 437–40. Winterville, N.C.: J. F. Morton.

TAMARIND

Dense evergreen tree, to 40 feet tall, with 2- to 8-inch-long, brownish-red, brittle-shelled fruit pods with thick, brown, sticky, sweet pulp surrounding a few seeds. The fruit is eaten fresh, flavors preserves and chutneys, and is added to meat sauces and juice.

Fisch, B. E. 1974. The tamarind. *CRFG Yearbook* 6:221–50.

Morton, J. F. 1987. *Fruits of warm climates*, 115–21. Winterville, N.C.: J. F. Morton.

TANGELO

Vigorous, fast-growing tree to 30 feet tall, with February to April ripening, tangerine-flavored fruit.

IC&RF Soc. 1988. The many mandarins (also tangelos and tangers). *IC&RF Soc. Newsl.* 27:10–12.

Morton, J. F. 1987. *Fruits of warm climates*, 158–60. Winterville, N.C.: J. F. Morton.

TANGERINE—See *Mandarin Orange*

TANGOR
Tangerine orange hybrid having a 6- to 12-foot-high and -wide, bushy, thorny tree with early spring-ripening, juicy, uniquely flavored fruit.

IC&RF Soc. 1988. The many mandarins (also tangelos and tangers). *IC&RF Soc. Newsl.* 27:10–12.

Morton, J. F. 1987. *Fruits of warm climates*, 145–46. Winterville, N.C.: J. F. Morton.

TARO

Violet, Stemmed (*Xanthosoma violacea*)
Harrington, G. 1978. *Grow your own Chinese vegetables*, 194–97. Pownal, Vt.: Garden Way Publishing.

Dasheen (*Colocasia antiquorum* or *C. esculeta*)
The leaves are cooked as a vegetable, with tubers ready from three to six months after setting out corms, depending on variety.

Chauhan, D. V. S. 1968. *Vegetable production in India*, 227–31. Agra: Ram Prasad & Sons.

TOMATILLO
Sprawling 1- to 2-foot-tall by 3- to 4-foot-wide annual with 1- to 2½-inch-diameter, green to yellow-green, sticky and sweet fruit that bursts through the husk when ripe and is used in salsa and jam.

Dremann, C. 1985. *Ground cherries, husk tomatoes, and tomatillos*, 1–2, 11–14. Redwood City, Calif.: Redwood City Seed Company.

Morton, J. F. 1987. Mexican husk tomatoes. *Fruits of warm climates*, 434–37. Winterville, N.C.: J. F. Morton.

TOMATO
Whealy, K. 1988. *The garden seed inventory*. Decorah, Ia.: Seed Saver Publications. For unusual nonhybrid varieties available in 1987. For hybrids and post-1987 nonhybrid varieties, browse general and "specialty" seed catalogs.

TROPICAL FRUIT—See *Fruits: Temperate*

UDO (*Aralia cordata*)
The blanched shoots of this perennial are cooked or used raw in Japan.

Fairchild, D. 1914. *Experiment with udo, the new Japanese vegetable*. USDA Bulletin 84.

VEGETABLES—Baby
Staff. 1986. What's with these little guys? *Sunset* 176(5):282–84.

Walker, S. L. 1986. Smallness for its own sake is now passé. *Calif. Farmer* 264(5):54–55.

WATER CHESTNUT
Sedge family plant that, in shallow water and muddy bottom, produces crisp, nutty-flavored, underwater nuts used raw in salads, in stir-fry dishes, and in egg roll stuffing.

McGeachin, R. B., and R. R. Stickney. 1979. *Culture of Chinese waterchestnuts in the southeastern U.S.* Proc. 33rd Ann. Conf. Southeastern Assoc. Fish & Wildlife Agencies, Nashville, Tenn., pp. 606–10.

Preacher, J. 1983. Backyard water chestnuts. *Plants Gard.* 39(2):39–41.

Twigg, B. A., F. C. Stark, and A. Kramer. 1957. Cultural studies with matai. *Am. Soc. Hort. Sci. Proc.* 70:266–72.

WATER CRESS
Shear, G. M. 1968. *Commercial growing of water cress*. USDA Farmers' Bulletin 2233.

WATER SPINACH
Annual, prostrate, herbaceous plant with smooth, extensively spreading vines. The young, 12-inch ends are edible and are grown either in shallow, muddy water or on heavily irrigated "dry" beds.

Rubatsky, V., and M. Yamaguchi, M. 1997. *World vegetables: Principles, production, and nutritive values*, (2nd ed.). New York.: Chapman and Hall Publishing Co.

Tisbe, V. O., and T. G. Cadiz. 1967. Convolvulus or kangkong. In J. E. Knott and J. R. Deanon (eds.), *Vegetable production in Southeast Asia*, 285–89. Los Baños: University of the Philippines College of Agriculture.

WAX GOURD—See *Gourd: Chinese winter melon*

WINGED BEAN—See *Beans: Winged*

YAMS
Degras, L., and F. Martin. 1978. *Tropical yams and their potential.* USDA Agriculture Handbooks Nos. 457, 466, 495, 502, 522.

Chinese (*Dioscorea batatas, D. esculenta*)
The tubers are a good source of vitamin C. The viney plants need to be staked for adequate yields. Table-ready by grating, boiling, and slicing.

Enyi, B. A. C. 1970. Growth studies in Chinese yams. In *Tropical Root and Tuber Crops Tomorrow, [Proc. Symp. Int. Soc. Trop. Root Crops.],* 103. Ottawa: International Development Research Center.

Kawakami, K. 1970. Yam culture in Hawaii. In *Tropical Root and Tuber Crops Tomorrow [Proc. Symp. Int. Soc. Trop. Root Crops.],* 102. Ottawa: International Development Research Center.

Okuli, O. O. 1983. *Rapid propagation of yam by the minisett technique.* In J. J. Cock (ed), Global Workshop on Root and Tuber Crops Propagation: Proceedings of a Regional Workshop, Cali, Columbia, Sept. 13–16, pp. 119–22. Whereas whole tubers or large tuber pieces are traditionally used for propagation, this describes a method for using much smaller pieces.

YARD-LONG BEANS—See *Beans: Yard-long*

Books for Your Shelf

California Agricultural Directory (including Oregon & Washington)
California Farm Bureau Federation. Annual publication (1601 Exposition Blvd., Sacramento, CA 95815). Contains addresses of county agricultural offices, state agricultural publications, and universities that teach agricultural subjects.

California Farm Fresh Directory
Community Alliance with Family Farmers (CAFF), P.O. Box 464, Davis, CA 95617.

Fruit, Berry, and Nut Inventory
Whealy, Kent (ed.). 1989. Decorah, Iowa: Seed Saver Publications (Rural Route 3, Box 239, Decorah, IA 52101). Presents the only complete listing of mail-order nurseries in the United States and Canada that are selling hybrid fruit, nut, and berry varieties. 366 pp.

The Garden Seed Inventory (2nd ed.)
Whealy, Kent. 1988. Decorah, Iowa: Seed Saver Publications (Rural Route 3, Box 239, Decorah, IA 52101). Presents the only complete listing of seed companies selling specific open-pollinated vegetable varieties.

Knott's Handbook for Vegetable Growers (3rd ed.)
Lorenz, O. A., and D. N. Maynard. 1988. New York: Wiley. 456 pp. Overly detailed but thorough.

National Organic Directory
Community Alliance with Family Farmers, P.O. Box 464, Davis, CA 95617). The most complete source of information about the North American organic food and farm supply market. It features listings for some 300 grower-wholesalers, distributors, and farm suppliers in the United States and Canada, as well as a comprehensive guide to certification programs, trade organizations, and publications, and timely articles on certification efforts throughout the nation.

Sunset New Western Garden Book (5th ed.)
Staff. 1988. Menlo Park, Calif.: Sunset Books and Sunset Magazine, Lane Publication. Its *Western Plant Encyclopedia* section presents detailed climate zones that correlate with most of the crops listed, along with some varieties.

Books from the Library

FRUITS

Fruits of India: Tropical and Sub-Tropical
Bose, T. K. 1985. Calcutta, India: Naya Prokash.

Fruits of Warm Climates
Morton, Julia F. 1987. Winterville, N.C.: J. F. Morton. Distributed by Creative Resources Systems.

The Handbook of Soft Fruit Growing
Turner, David, and Ken Muir. 1985. London and Dover, N.H.: C. Helm. For the United Kingdom, but good if used with Sunset's *New Western Garden Book*. Soft fruits are berries.

Modern Fruit Science
Childers, Norman F. 1983. Gainesville, Fla.: Horticultural Pub.

Uncommon Fruits & Vegetables: A Common-Sense Guide
Schneider, Elizabeth. 1986. New York: Harper & Row. A cookbook with a vast array of the unusual, with very useful information on choosing quality produce and storage for each product.

Wild Fruits of the Sub-Himalayan Region
Parmar, C., and M. K. Kaushal. 1982. Ludhiana, India: Kalyani Pub. For purchasing details and ways to obtain seed, contact Dr. Chiranjit Parmar, Sharda Niwas (Near Ceeta Ashram), Thodo Ground, Solan, H.P. 173 212 INDIA. Based on visits to California, Dr. Parmar states that the climate of the Himalayan mid-hills is somewhat similar to that of parts of California. The book covers 26 wild fruits in fair detail and includes judgments on their quality and potential uses.

HERBS
Growing and Using Herbs and Spices
Miloradovich, Milo. 1986. New York: Dover.

Herbs: An Indexed Bibliography, 1971–1980: The Scientific Literature on Selected Herbs, and Aromatic and Medicinal Plants of the Temperate Zone
Simon, James E. 1984. Hamden, Conn.: Archon Books. A good reference source, but difficult to find useful citations.

Herbs: How to Select, Grow, and Enjoy
Lathrop, Norma Jean. 1981. Tucson, Ariz.: H-P Books. Covers only a few herbs that *Park's Success with Herbs* doesn't have, and the growing instructions are minimal.

Park's Success with Herbs
Foster, Gertrude B. 1980. Greenwood, S.C.: G.W. Park Seed Co. The best herb book I've found for the cultivation of a wide variety of herbs.

Rodale's Illustrated Encyclopedia of Herbs
Kowalchik, C., and W. H. Hylton (eds.). 1987. Emmaus, Penn.: Rodale Press. Slightly more complete than *Park's Success with Herbs* but probably better used as an adjunct.

MUSHROOMS
Growing Shiitake Commercially: A Practical Manual for the Production of Japanese Forest Mushrooms
Harris, Bob. 1986. Madison, Wis.: Science Tech Pubs. 72 pp.

The Mushroom Cultivator: A Practical Guide to Growing Mushrooms at Home
Stamets, Paul, and J. S. Chilton. 1983. Olympia, Wash.: Agarikon Press. Very complete and detailed, with the exception of truffles.

NUTS

Nut Tree Culture in North America
Jaynes, R. A. (ed). 1979. Hamden, Conn.: North America Nut Growers Assoc.

VEGETABLES

The Forgotten Art of Flower Cookery
Smith, Leona S. 1973. New York: Harper & Row. A recipe book with a good rundown of edible flowers.

Ground Cherries, Husk Tomatoes, and Tomatillos
Dremann, Craig. 1985. Redwood City, Calif.: Redwood City Seed Company. Considered the best growing directions in the United States. 22 pp.

Grow Your Own Chinese Vegetables
Harrington, Geri. 1978. Pownal, Vt.: Garden Way Publishing; reprinted 1985. 298 pp.

Manual of Minor Vegetables
Stephens, James M. 1988. Gainesville, Fla.: University of Florida (Florida Cooperative Extension, Bulletin SP-40, University of Florida, Gainesville, FL 32611). Lists a few crops not covered elsewhere with rather rudimentary cultivation information. Use in conjunction with Sunset climatic zone guides.

Onions and Their Allies: Botany, Cultivation, and Utilization
Jones, Henry Albert, and Louis K. Mann. 1963. London: L. Hill; New York: Interscience Publishers.

The Salad Garden: Salads from Seed to Table
Larkcom, Joy. 1984. New York: Viking Press. A fairly complete listing of the vast number of American and European salad ingredients, but provides little rudimentary cultivation instruction.

Vegetable Production in India (2nd ed.)
Chauhan, D. V. S. 1968. Agra: Ram Prasad & Sons. Has many unusual crops with growing instructions for India.

Vegetable Production in Southeast Asia
Knott, James E., and Jose R. Deanon (eds.). 1967. Los Baños: University of the Philippines College of Agriculture. 366 pp.

World Vegetables: Principles, Production, and Nutritive Values (2nd ed.)
Rubatsky, Vincent, and Mas Yamaguchi. 1997. New York: Chapman and Hall Publishing Co.

GENERAL

Tropical Products Transport Handbook
U.S. Department of Agriculture. 1989. Agricultural Handbook 668. Advice on maintaining quality of cut flowers, plants, fruit, and vegetables in warmer climates and information on USDA requirements for improving pest-free foreign produce.

Note: The best place to obtain agricultural books in northern California is agAccess, P.O. Box 2008, Davis, CA 95617. Telephone (530) 756-7177; Fax (530) 756-7188.

Useful Free Publications

California Agriculture
(Monthly)
University of California
Division of Agriculture and Natural Resources
300 Lakeside Drive, 6th floor
Oakland, CA 94612-3560

Good Bug Directory
The fourth edition of *Suppliers of Beneficial Organisms in North America* lists 60 companies that sell biological controls. Among the 60 organisms available are predatory mites, fly parasites, parasitic nematodes, and other organisms to help control pest invasions in orchards, ranches, greenhouses, farms, and ornamental gardens. Available from: California Department of Food and Agriculture, Biological Control Services Program, 3288 Meadowview Road, Sacramento, CA 95832

Local County Cooperative Extension Service Publications
See *UC DANR Communication Services Catalog* for the address and telephone number of your county agent who will provide you with a list of publications.

Small Farm News
(Bi-monthly; list of free publications is also available)
Small Farm Center
One Shields Avenue
University of California
Davis, CA 95616-8699

Sustainable Agriculture News
(Monthly)
UC Sustainable Agriculture Research and Education Program
One Shields Avenue
University of California
Davis, CA 95616

UC DANR Communication Services Catalog
(Annual)
Special Publication 3020
University of California
Division of Agriculture and Natural Resources
Communication Services
6701 San Pablo Avenue, 2nd Floor
Oakland, CA 94608-1239
(phone 1-800-994-8849)

Recommended Journals

Business of Herbs
The international news and resource service on herb businesses. P.O. Box 559, Madison, VA 22727; telephone (703) 948-7169. Bimonthly.

California Farmer
ABC Cap Cities, 2300 Clayton Road, #1360, Concord, CA 94520. Usually at least one item of interest to smaller growers and articles to help you keep up with the "other" California agriculture. 15 issues annually.

CRFG (California Rare Fruit Growers)
Newsletter, Fruit Gardener, Journal, Yearbook
(Took over where IC&RF left off)
Fruit Gardener
Fullerton Arboretum, CSUF, Fullerton, CA 92634. Provides contacts with members in your area as well as throughout the state, along with individuals who are experts. Newsletter (1969–86), Fruit Gardener (1987–), Journal (1986–) Yearbook (1969–85). Quarterly.

C.C.O.F. Statewide Newsletter
California Certified Organic Farmers, Box 8136, Santa Cruz, CA 95061.

Fine Gardening
The Taunton Press, Newtown, CT 06470; telephone 1-800-243-7252. Usually has one or two articles a month on fairly uncommon fruit and vegetable crops written by people who are growing them.

Fruit Gardener—See CRFG

Herb, Spice, and Medicinal Plant Digest
Department of Plant and Soil Sciences Cooperative Extension, Stockbridge Hall, University of Massachusetts, Amhurst, MA 01003. Quarterly.

IPM Practitioner
Bio-Integral Resource Center (BIRC), P.O. Box 7414, Berkeley, CA 94707. This monthly journal usually features one aspect of IPM and summarizes other areas. Membership also includes inquiry privileges and assistance in pest-control information searches.

Organic Gardening
33 E. Minor St., Emmaus, PA 18098.

Organic Growers List and Crop Index
CCOF, P.O. Box 8136, Santa Cruz, CA 95061. The annual CCOF directory of organically grown crops, certified growers, growers who ship direct, and farms offering apprenticeships. Yearly.

Pomona
(NAFEX) NAFEX, Rt. 1, Box 94, Chapon, IL 62628. This nationwide organization deals chiefly with old-fashioned or odd varieties of temperate fruits. Quarterly.

Seed Saving Project News
Seed Saving Project, Department LAWR, 139 Hoagland Hall, Davis, CA 95616. Dedicated to maintaining and distributing seeds of rare and endangered vegetable, flower, and herb varieties by exchanging seed via the newsletter and in person.

Sunset Magazine
Lane Publishing, Menlo Park, CA 94025. It is not worth spending the money for the few helpful articles but worth looking through at the local library every month for articles on new foods and occasionally new crops.

Tropical Fruit News
Rare Fruit Council International, Inc., P.O. Box 561914, Miami, FL 33256. This monthly eight-page journal deals with more tropical unusual fruits and nuts.

Seed Sources
NATIONAL

Abundant Life Seed Foundation
P.O. Box 772
Port Townsend, WA 98368
(Catalog and newsletters. Good source of open-pollinates and only source for quinoa and some grain amaranths.)

Johnny's Selected Seeds
Foss Hill Road
Albion, ME 04910
(Catalog. Gives good cultivation, adaptability information. Request Growers' Catalog.)

Park Seed Company
Cokesbury Road
Greenwood, SC 29647-0001
(Free catalog; separate retail and wholesale catalogs.)

Seed World (Seed Trade, Buyer's Guide)
380 Northwest Highway
Des Plaines, IL 60016
(Annually every April. Seed World also publishes an annual directory of seed companies in the United States.)

Thompson & Morgan, Inc.
P.O. Box 1308
Jackson, NJ 08527
(Free catalog. Enormous variety, especially strong in "common" European vegetables and varieties.)

CALIFORNIA

Fungi

Far West Fungi
P.O. Box 428
South San Francisco, CA 94083

Mushroom People
P.O. Box 159-A
Inverness, CA 94937
(415) 663-8504

General

Bountiful Gardens
19550 Walker Road
Willits, CA 95490

Clyde Robin Seed Co.
25670 Nickel Pl.
Hayward, CA 94545
(510) 785-0425

Mark Seed and Supply Co.
3401 33rd Avenue
Sacramento, CA 95824
(916) 421-7334
(916) 421-SEED

Ornamental Edibles
3622 Weedin Court
San Jose, CA 95132

(The) Redwood Seed Co.
P.O. Box 361
Redwood City, CA 94064
(650) 325-7333
(Request bulk seed catalog.)

Taylor's Herb Gardens, Inc.
1535 Lone Oak Road
Vista, CA 92084
(916) 727-3485

Van Ness Water Gardens
2460 North Euclid Avenue
Upland, CA 91786

Oriental

Bonanza Seeds International
P.O. Box V
Gilroy, CA 95020
(Wholesale only.)

Harris Moran Seeds, Inc.
1155 Harkins Rd.
Salinas, CA 93901

Kitazawa Seed Co.
1748 Laine Avenue
Santa Clara, CA 95051

Neuman Seeds, Inc.
P.O. Box 1373
El Centro, CA 92244

Tsang and Ma
P.O. Box 294
Belmont, CA 94002
(650) 595-2270
(Request wholesale catalog.)

NURSERIES

Fowler Nurseries, Inc.
525 Fowler Road
Newcastle, CA 95658
(916) 645-8191 - Catalog $1

Living Tree Center
P.O. Box 797
Bolinas, CA 94924
(415) 868-2224 - Journal/Catalog $6, deductible from first order

Peaceful Valley Farm Supply
11173 Peaceful Valley Road
Nevada City, CA 95959
(530) 265-3276 - Catalog $2, deductible from first order

Sonoma Antique Apple Nursery
4395 Westside Road
Healdsburg, CA 95448

INDEX TO SCIENTIFIC AND COMMON NAMES

A
Abelmoschus esculentus, 89
adzuki bean, 5
African horned cucumber, 74
African horned melon, 74
aka takana, 114
Alice, sweet, 7
Allium ampeloprasum, 77
Allium fistulosum, 69
Allium schoenoprasum, 43
Allium tuberosum, 65
amaranth, vegetable, 134
Amaranthus spp., 134
American black currant, 56
ampalaya, 22
Anethum graveolens, 52
anise, 7
anise, sweet, 60
apple pear, 11
Artemisia dracunculus, 125
arugula, 9
ash gourd, 138
Asian pear, 11
asparagus bean, 41
asparagus lettuce, 36
azuki, 5

B
baby corn, 15
balatung, 84
balsam pear, 22
basil, 17
Batavian lettuce, 112
bau, 27
bean, adzuki, 5
bean, asparagus, 41
bean, bell, 58
bean, broad, 58
bean, English dwarf, 58
bean, fava, 58
bean, feve, 58
bean, haba, 58
bean, horse, 58
bean, long, 41
bean, mung, 84
bean, pigeon, 58

bean, silkworm, 58
bean, tick, 58
bean, Windsor, 58
bean, yam, 71
beet, leaf, 123
beet, spinach, 123
beet, Swiss, 123
Belgian endive, 20
bell bean, 58
Benincasa hispida, 138
Beta vulgaris, 123
bibb lettuce, 112
bitter gourd, 22
bitter melon, 22
black currant, American, 56
black currant, European, 56
bok choy, 25
bottle gourd, 27
Brassica campestris, 29
Brassica juncea, 114
Brassica napus, 29
Brassica oleracea, 39, 48, 75
Brassica rapa, 25, 86, 132
broad bean, 58
broccoli, Chinese, 39
bunching onion, Japanese, 69
butterhead lettuce, 112

C
cabatiti, 120
cabbage, Chinese, 86
cabbage, nappa, 86
cabbage, swamp, 136
cactus pear, 94
calabash gourd, 27
callaloo, 134
canola, 29
caper, 32
Capparis spinosa, 32
cardoon, 34
celery cabbage, 86
celtuce, 36
cham kwa, 138
chard, 123
chayote, 37
Chenopodium quinoa, 98

chicon, 20
chicory, 54
chicory, red, 100
chicory, witloof, 20
chiifu, 86
Chinese broccoli, 39
Chinese cabbage, 86
Chinese chives, 65
Chinese green mustard, 114
Chinese kale, 39
Chinese long bean, 41
Chinese okra, 120
Chinese parsley, 45
Chinese pear, 11
Chinese radish, 50
Chinese water spinach, 136
ching quat, 136
chives, 43
chives, Chinese, 65
chives, garlic, 65
cho cho, 37
cholai bhaji, 134
Christmas melon, 138
chung, 69
Cichorium endivia, 54
Cichorium intybus, 20, 100
cilantro, 45
cinnamon basil, 17
citron, 47
citronella grass, 80
Citrullus lanatus, 47
Citrullus vulgaris, 47
cladodes, 94
clove currants, 56
collards, 48
common fennel, 60
cong, 69
convolvulus, water, 136
coriander, 45
Coriandrum sativum, 45
corn, baby, 15
cos lettuce, 112
creeping golden marjoram, 82
cu-cai trang, 50
cucumber, African horned, 74
Cucumis metuliferus, 74

cucuzzi, 27
currant, American black, 56
currant, clove, 56
currant, European black, 56
currant, red, 102
currant, white, 102
Cymbopogon citratus, 80
Cymbopogon nardus, 80
Cynara cardunculus, 34

D
dai gai choy, 114
daikon, 50
dau-dua, 41
dau-xanh, 84
dill, 52
dishrag gourd, 120
doongua, 138
dow gauk, 41

E
endive, 54
endive, Belgian, 20
endive, French, 20
English dwarf bean, 58
Eruca sativa, 9
escarole, 54
European black currant, 56

F
fava bean, 58
fennel, 60
feve bean, 58
Ficus carica, 63
fig, fresh, 63
fig, Indian, 94
figadindi, 94
finocchio, 60
Florence fennel, 60
Foeniculum officinale, 60
Foeniculum vulgare, 60
French endive, 20
fu kwa, 22
fut shau kua, 37

G
gai choy, 114
gai-lohn, 39
garlic chives, 65
gil choy, 65
golden marjoram, 82
gooseberry, 67
gourd, ash, 138

gourd, bitter, 22
gourd, bottle, 27
gourd, calabash, 27
gourd, sponge, 120
gourd, wax, 138
gow choy, 65
grass, citronella, 80
Greek oregano, 91
gumbo, 89
gwa tsz tsai, 97

H
haba bean, 58
hakusai, 86
hanh-ta, 69
hayato uri, 37
he, 43
hechima, 120
hinn choy, 134
horned cucumber, 74
horned melon, 74
horse bean, 58
hsien tsai, 134

I
India mustard, 114
Indian fig, 94
Ipomoea aquatica, 136
Ipomoea raptans, 136

J
Japanese bunching onion, 69
Japanese pear, 11
jelly melon, 74
jicama, 71
jiu tsai, 65
joh tsu, 45
juro-kusasagemae, 41

K
kailan, 39
kale, Chinese, 39
kalunay, 134
kang kong, 136
kang kung, 136
kankon, 136
kau tsai, 65
kerala, 22
kinchi, 45
kiwano, 74
knot marjoram, 82
koendoro, 45
kohlrabi, 75

kolo taac, 84
kui tsai, 65
kulitis, 134
kwoon taat tsoi, 123
kyona, 114

L
labanos, 50
Lactuca sativa, 36, 112
Lagenaria siceraria, 27
lal sag, 134
leaf beet, 123
leaf lettuce, 112
leek, 77
lemon basil, 17
lemongrass, 80
lettuce, 112
lettuce, asparagus, 36
Lippia graveolens, 91
lobok, 50
long bean, 41
loofah, 120
look dou, 84
lo pue, 50
lor bark, 50
lu tou, 84
luffa, 120
Luffa acutangula, 120
Luffa cylindrica, 120
Lycopersicon lycopersicum, 116

M
Majorana hortensis, 82
marjoram, 82
marjoram, creeping golden, 82
marjoram, knot, 82
marjoram, pot, 91
marjoram, sweet, 82
marjoram, wild, 91
marjoram, winter, 91
melon, African horned, 74
melon, bitter, 22
melon, Christmas, 138
melon, jelly, 74
melon, preserving, 47
melon, winter, 138
Mexican oregano, 91
Mexican parsley, 45
mirliton, 37
mizuna, 114
Momordica charantia, 22
moyashi, 84
moyashi-mame, 84

multiplier onion, 69
mung bean, 84
muop khia, 120
mustard, Chinese green, 114
mustard, potherb, 114

N
nappa cabbage, 86
nashi, 11
ndoh dah, 65
ndoh trah, 69
nebuka, 69
negi, 69
New Zealand spinach, 88
ngalog, 97
ngar choy, 84
ngow-lai choi, 37
nigai uri, 22
nihonnashi, 11
nira, 65
nong taao, 84
nopales, nopalitos, 94

O
Ocimum spp., 17
oilseed rape, 29
okra, 89
okra, Chinese, 120
ong tsoi, 136
onion, Japanese bunching, 69
onion, multiplier, 69
onion, Welsh, 69
Opuntia spp., 94
oregano, 91
Oriental pear, 11
Oriental radish, 50
Origanum majorana, 82
Origanum vulgare, 91
oyster plant, 110

P
paak tim tsoi, 123
Pachyrrhizus erosus, 71
pai-tsai, 86
pak bung, 136
pak kah nah, 39
pak quat, 136
parsley, Chinese, 45
parsley, Mexican, 45
parsnip, 93
Pastinaca sativa, 93
patola, 120
pear, apple, 11

pear, Asian, 11
pear, balsam, 22
pear, cactus, 94
pear, prickly, 94
pear, vegetable, 37
pechay, 86
pe-tsai, 86
Peurto Rican oregano, 91
Phaseolus angularis, 5
Physalis philadelphica, 130
pigeon bean, 58
Pimpinella anisum, 7
po gua, 27
Poliomintha longiflora, 91
Portulaca oleracea, 97
pot marjoram, 91
potherb mustard, 114
pourpier, 97
preserving melon, 47
prickly pear, 94
pua sha, 84
purslane, 97
Pyrus serotina, 11

Q
quinoa, 98

R
radicchio, 100
radish, Chinese, 50
radish, Oriental, 50
rao mui, 45
rape, oilseed, 29
Raphanus sativus, 50
rau muong, 136
red chicory, 100
red currant, 102
Ribes americanum, 56
Ribes hirtellum, 67
Ribes nigrum, 56
Ribes odoratum, 56
Ribes petraeum, 102
Ribes rubrum, 102
Ribes sativum, 102
Ribes uva-crispa, 67
Ribes vulgare, 102
rocket salad, 9
romaine lettuce, 112
roquette, 9
rosemary, 104
Rosmarinus officinalis, 104

S
sage, 107
salad pear, 11
salsify, 110
Salvia spp., 107
sayote, 37
Scolymus hispanicus, 110
Scorzonera hispanica, 110
Sechium edule, 37
sibuyas, 69
silkworm bean, 58
sinqua, 120
sitaw, 41
siu heung, 43
skoo ah, 120
smooth gourd, 120
specialty lettuces, 112
specialty mustards, 114
speciatly tomatoes, 116
spinach beet, 123
spinach, Chinese water, 136
spinach, New Zealand, 88
sponge gourd, 120
stem turnip, 75
swamp cabbage, 136
sweet Alice, 7
sweet anise, 60
sweet basil, 17
sweet fennel, 60
sweet marjoram, 82
Swiss beet, 123
Swiss chard, 123

T
taao-hla-chao, 41
tampala, 134
tankoy, 138
tao tah, 37
tarragon, 125
ta tsu kua, 120
Tetragonia tetragonioides, 88
thyme, 127
Thymus spp., 127
tick bean, 58
to jisa, 123
tomatillo, 130
tomato, 116
toongsin tsai, 136
tougan, 138
Tragopogon porrifolius, 110
trai su, 37
tsai hsio li, 37
tsina, 86

tsung, 43
tunas, 94
turnip, 132
turnip, stem, 75

U
upo, 27

V
vegetable amaranth, 134
vegetable pear, 37
verdolaga, 97
Vicia faba, 58
Vigna radiata, 84
Vigna unguiculata subsp. *sesquipedalis,* 41

W
water convolvulus, 136
water spinach, Chinese, 136
wax gourd, 138
Welsh onion, 69
weng cai, 136
white currant, 102
wild fennel, 60
wild marjoram, 91
Windsor bean, 58
winter marjoram, 91
winter melon, 138
witloof chicory, 20
won bok, 86

X
xu-xu, 37

Y
yam bean, 71
yar tsai, 84
yard-long bean, 41
yeung poh tsoi, 88
yim sai, 45
you-sai, 136
yuan sui, 45
yugao, 27
yun tsai, 45

Z
Zea mays, 15